Spatial
Microsimulation
with R

Chapman & Hall/CRC
The R Series

Series Editors

John M. Chambers
Department of Statistics
Stanford University
Stanford, California, USA

Torsten Hothorn
Division of Biostatistics
University of Zurich
Switzerland

Duncan Temple Lang
Department of Statistics
University of California, Davis
Davis, California, USA

Hadley Wickham
RStudio
Boston, Massachusetts, USA

Aims and Scope

This book series reflects the recent rapid growth in the development and application of R, the programming language and software environment for statistical computing and graphics. R is now widely used in academic research, education, and industry. It is constantly growing, with new versions of the core software released regularly and more than 7,000 packages available. It is difficult for the documentation to keep pace with the expansion of the software, and this vital book series provides a forum for the publication of books covering many aspects of the development and application of R.

The scope of the series is wide, covering three main threads:

- Applications of R to specific disciplines such as biology, epidemiology, genetics, engineering, finance, and the social sciences.
- Using R for the study of topics of statistical methodology, such as linear and mixed modeling, time series, Bayesian methods, and missing data.
- The development of R, including programming, building packages, and graphics.

The books will appeal to programmers and developers of R software, as well as applied statisticians and data analysts in many fields. The books will feature detailed worked examples and R code fully integrated into the text, ensuring their usefulness to researchers, practitioners and students.

Published Titles

Spatial Microsimulation with R, *Robin Lovelace and Morgane Dumont*

Statistics in Toxicology Using R, *Ludwig A. Hothorn*

Stated Preference Methods Using R, *Hideo Aizaki, Tomoaki Nakatani, and Kazuo Sato*

Using R for Numerical Analysis in Science and Engineering, *Victor A. Bloomfield*

Event History Analysis with R, *Göran Broström*

Computational Actuarial Science with R, *Arthur Charpentier*

Statistical Computing in C++ and R, *Randall L. Eubank and Ana Kupresanin*

Basics of Matrix Algebra for Statistics with R, *Nick Fieller*

Reproducible Research with R and RStudio, Second Edition, *Christopher Gandrud*

R and MATLAB®, *David E. Hiebeler*

Nonparametric Statistical Methods Using R, *John Kloke and Joseph McKean*

Displaying Time Series, Spatial, and Space-Time Data with R, *Oscar Perpiñán Lamigueiro*

Programming Graphical User Interfaces with R, *Michael F. Lawrence and John Verzani*

Analyzing Sensory Data with R, *Sébastien Lê and Theirry Worch*

Parallel Computing for Data Science: With Examples in R, C++ and CUDA, *Norman Matloff*

Analyzing Baseball Data with R, *Max Marchi and Jim Albert*

Growth Curve Analysis and Visualization Using R, *Daniel Mirman*

R Graphics, Second Edition, *Paul Murrell*

Introductory Fisheries Analyses with R, *Derek H. Ogle*

Data Science in R: A Case Studies Approach to Computational Reasoning and Problem Solving, *Deborah Nolan and Duncan Temple Lang*

Multiple Factor Analysis by Example Using R, *Jérôme Pagès*

Customer and Business Analytics: Applied Data Mining for Business Decision Making Using R, *Daniel S. Putler and Robert E. Krider*

Implementing Reproducible Research, *Victoria Stodden, Friedrich Leisch, and Roger D. Peng*

Graphical Data Analysis with R, *Antony Unwin*

Using R for Introductory Statistics, Second Edition, *John Verzani*

Advanced R, *Hadley Wickham*

Dynamic Documents with R and knitr, Second Edition, *Yihui Xie*

Spatial Microsimulation with R

Robin Lovelace
Morgane Dumont

With the assistance of
Richard Ellison and Maja Založnik

CRC Press
Taylor & Francis Group
Boca Raton London New York

CRC Press is an imprint of the
Taylor & Francis Group, an **informa** business

A CHAPMAN & HALL BOOK

CRC Press
Taylor & Francis Group
6000 Broken Sound Parkway NW, Suite 300
Boca Raton, FL 33487-2742

© 2016 by Taylor & Francis Group, LLC
CRC Press is an imprint of Taylor & Francis Group, an Informa business

No claim to original U.S. Government works

Printed on acid-free paper
Version Date: 20160125

International Standard Book Number-13: 978-1-4987-1154-8 (Paperback)

Visit the Taylor & Francis Web site at
http://www.taylorandfrancis.com

and the CRC Press Web site at
http://www.crcpress.com

Contents

Preface

Spatial microsimulation is a set of methods for modelling phenomena which operate at individual and geographical levels simultaneously. For example the functioning of a modern city (Shanghai is illustrated on the front cover) involves an overwhelmingly complex web of human interactions. Simulating such complexity may seem impossible. Yet, by breaking the problem up into its constituent parts — discrete geographic areas and a sample of the population — spatial microsimulation can be used to model key aspects of the city system on an everyday laptop computer. There are dangers associated with reductionism, but condensing a problem down to its fundamentals has many advantages. Using the techniques described we can simulate scenarios such as population growth, increased energy efficiency and major shifts in transport technologies and modes. By linking the synthetic population to an agent-based model, in which individuals interact over time with each other and their environment, complex behaviours such as social segregation could also be simulated, as illustrated in Chapter 12.

This book is for anyone who wants to not only understand but to *use* spatial microsimulation. The emphasis is on the practical rather than theoretical aspects of the field. R packages such as **mipfp**, for enabling the generation of synthetic populations, are described in detail, with reference to practical examples and reproducible code. The aim is to enable you to implement the methods on your own data.

By explaining how to use tools for modelling phenomena that vary over space, this book should help enhance your knowledge of complex systems. We hope the book empowers the reader with the confidence and know-how needed to provide evidence-based policy guidance.

The origins of this book are more prosaic: during my PhD at the University of Sheffield I was tasked with using spatial microsimulation to model transport energy use. Despite the growing academic literature on the subject, there was little information that explained *how* to do spatial microsimulation, using a modern programming language. It was informal communication and code-sharing with a colleague, Malcolm Campbell, that led to the development of my models in R. This experience demonstrated the importance of reproducible research. Following this 'open science' ethic, readers are encouraged to comment on and contribute to the book's continued development via the code sharing site GitHub (see `https://github.com/Robinlovelace/spatial-microsim-book`).

The opportunity to turn the idea into reality came in the spring of 2014, when I developed notes for an 'Introduction to Spatial Microsimulation' course at the University of Leeds. The high demand for and positive feedback after the course suggested the need for practical teaching materials in the area. Four months later I delivered another course on spatial microsimulation at the University of Cambridge. The materials had been greatly updated and, thanks to the involvement of CRC Press, these provided the foundation for a book on the subject.

Morgane Dumont (NaXys, University of Namur), who attended the Cambridge course, became involved shortly after and has greatly improved the work. Morgane's background in Mathematics and Statistics made her the ideal co-author, complementing the focus on practical examples and code.

Maja Zaloznik (University of Oxford) and Richard Ellison (University of Sydney) have also greatly improved the book through their contributed chapters. Richard's chapter (11) illustrates how R can be used as the basis for transport demand modelling, using an approach known as TRESIS. Maja's chapter (12) is the most advanced in the book and demonstrates how spatial microsimulation can be used in parallel with agent-based modelling, with an implementation in the NetLogo language.

Spatial microsimulation with R is therefore the result of international teamwork. It is, to the best of our knowledge, the only practical book on the subject. We hope it is useful in your work. More widely, we hope it provides a solid foundation for advancement in the field and a toolkit for solving real-world problems.

If you have any feedback on the book please do get in touch via the book's online repository, hosted on the code sharing platform GitHub: https://github.com/Robinlovelace/spatial-microsim-book.

Robin Lovelace, February 2016.

Acknowledgements

As with any worthwhile textbook, this was not a solo effort. We benefited immensely from teaching spatial microsimulation to diverse audiences, the formal and informal feedback they provided, and correspondence with a number of people using spatial microsimulation 'in the wild'. Of these, the following deserve special mention:

- James Gleeson, from the Greater London Authority (GLA), provided insight into how spatial microsimulation can be used in local government and made several improvements to the book.

- Ulrike Rauer, from the University of Oxford, commented on early drafts of the book and showed how it could be made more relevant to PhD students new to the approach.

- Stephen Clarke at the University of Leeds demonstrated the benefits of the Flexible Modelling Framework and encouraged testing of the R code on much larger datasets than had previously been used, encouraging optimisation of the code.

- Johan Barthélemy, from the SMART Infrastructure (Wollongong), helped in understanding his methods and R package (mipfp).

- Lex Comber, who provided crucial comments on the structure of the first part of the book and a great insight into how to make it more useful for teaching.

- Malcolm Campbell, my predecessor in the PhD. Malcolm provided a huge amount of support during the early phase of my PhD and shared all the R code he developed. He's been a great support of the book from the beginning.

- Everyone who provided input from the University of Leeds, including Mark Birkin, Nick Malleson and Andy Evans.

Morgane Dumont's research is funded by the Wallonia Region (Belgium) and she is a member of NaXys (University of Namur). She thanks these two affiliations in particular. We also thank the FNRS (Belgium) for the funding that allowed Morgane to meet Robin in Leeds in 2015 to work together on the book. Finally, computational resources (for Chapter 10) have been provided by the Consortium des Équipements de Calcul Intensif (CÉCI), funded by the Fonds de la Recherche Scientifique de Belgique (F.R.S.-FNRS) under Grant No. 2.5020.11.

Thanks also to all the people who provided the wider resources for this project to happen.

List of Figures

List of Tables

Part I

Introducing spatial
microsimulation with R

1

Introduction

CONTENTS

1.1 Who this book is for and how to use it

This book is for anyone who wants to learn how to *do* spatial microsimulation. By that we mean taking individual level and geographical datasets and processing them to solve real-world problems.

This book is a general-purpose introduction to the field and its implementation in a modern programming language. As such there is no single target audience, although the book was written with the following people in mind:

- Practitioners: because of the applied nature of the method and its utility for modelling people, it is useful in a wide range of sectors. Planning for the future of educational, health and transport services at the local level, for example, could benefit from spatial microdata on the inhabitants of the study area. The future is of course uncertain and spatial microsimulation can provide tools for scenario-based planning to explore various visions of the future. In this context spatial microsimulation can be especially useful for designing local policies to assist with the global transition away from fossil fuels (Lovelace and Philips, 2014).

- Academics: spatial microsimulation originated in academic research, building on the work of early pioneers such as Deming and Stephan (1940) and Clarke and Holm (1987). With recent advances in computer power, open source software and data accessibility, spatial microsimulation can now answer a greater range of questions than ever before. Because spatial microsimulation and associated methods are still in their infancy, there are still many areas of research where the approach has yet to be used. This makes spatial microsimulation an ideal method for doing new research, for example as part of a PhD.

- Educators: although spatial microsimulation has become more accessible over the last few years, it is still out of reach for the majority of people. Yet the potential benefits are large. This book provides example data and code that can be used as part of a course involving the analysis of spatial microdata, for example in a module contributing to an undergraduate or MSc course on Transport Modelling, Spatial Economics or Quantitative Geography.

The book has a definite progression from easier content to more advanced topics, especially those in Chapters 11 and 12, which deal with spatial microsimulation for transport modelling and agent-based modelling (ABM), respectively. The book can be read in order from front to back for a detailed overview of the field, its applications and its concepts (Part I); the practicalities and software decisions involved in generating spatial microdata (Part II); and issues around the modelling of spatial microdata (Part III).

Equally, the book can be used as a reference volume, to be dipped into as and when particular topics arise. There is a path dependency in the book however: Part II assumes that the reader is familiar with the concepts introduced in Part I, and the two chapters in Part III assume that the reader is competent with generating and handling spatial microdata, as introduced in Chapters 4 and 5. These are the central chapters in the book, in terms of generating spatial microdata with R. Chapters 4 and 5 are also the most important for people who simply want to generate spatial microdata rapidly. Chapter 6 provides insight into more experimental approaches for generating spatial microdata, while Chapter 7 provides a comprehensive worked example of the methods used 'in the wild'. A more detailed overview is provided in Section 1.8.

If you can't wait to get your hands dirty with code and example data for doing spatial microsimulation, please skip to Chapter 3. For other readers who want a little more background on this book and spatial microsimulation, bear with us. The remainder of this chapter explains the thinking behind the book, including the reasons for focussing on spatial microsimulation in R rather than in another language and the motivations for doing spatial microsimulation in the first place. The next chapter is practical. Chapter 3, by contrast, brings the reader up-to-date on what spatial microsimulation is and how it is currently used.

For the structure, each chapter begins with an introduction and ends with a summary of what has been done in the chapter. The last section of this chapter contains an overview of the book.

1.2 Motivations

Imagine a world in which data on companies, households and governments were widely available. Imagine, further, that researchers and decision-makers acting in the public interest had tools enabling them to test and *model* such data to explore different scenarios of the future. People would be able to make more informed decisions, based on the best available evidence. In this technocratic dreamland pressing problems such as climate change, inequality and poor human health could be solved.

These are the types of real-world issues that we hope the methods in this book will help to address. Spatial microsimulation can provide new insights into complex problems and, ultimately, lead to better decision-making. By shedding new light on existing information, the methods can help shift decision-making processes away from ideological bias and towards evidence-based policy.

The 'open data' movement has made many datasets more widely available. However, the dream sketched in the opening paragraph is still far from reality. Researchers typically must work with data that is incomplete or inaccessible. Available datasets often lack the spatial or temporal resolution required to understand complex processes. Publicly available datasets frequently miss key attributes, such as income. Even when high quality data is made available, it can be very difficult for others to check or *reproduce* results based on them. Strict conditions inhibiting data access and use are aimed at protecting citizen privacy but can also serve to block democratic and enlightened decision making.

The empowering potential of new information is encapsulated in the saying that 'knowledge is power'. This helps explain why methods such as spatial microsimulation, that help represent the full complexity of reality, are in high demand.

Spatial microsimulation is a growing approach to studying complex issues in the social sciences. It has been used extensively in fields as diverse as transport, health and education (see Chapter 3), and many more applications are possible. Fundamental to the approach are approximations of individual level data at high spatial resolution: people allocated to places. This *spatial microdata*, in one form or another, provides the basis for all spatial microsimulation research.

The purpose of this book is to teach methods for *doing* (not reading about!) spatial microsimulation. This involves techniques for generating and analysing spatial microdata to get the 'best of both worlds' from real individual and geographically-aggregated data. *Population synthesis* is therefore a key stage in spatial microsimulation: generally real spatial microdata are unavailable due to concerns over data privacy. Typically, synthetic spatial microdatasets are generated by combining aggregated outputs from Census results with individual level data (with little or no geographical information) from surveys that are representative of the population of interest.

The resulting *spatial microdata* are useful in many situations where individual level and geographically specific processes are in operation. Spatial microsimulation enables modelling and analysis on multiple levels. Spatial microsimulation also overlaps with (and provides useful initial conditions for) agent-based models (see Chapter 12).

Despite its utility, spatial microsimulation is little known outside the fields of human geography and regional science. The methods taught in this book have the potential to be useful in a wide range of applications. Spatial microsimulation has great potential to be applied to new areas for informing public policy. Work of great potential social benefit is already being done using spatial microsimulation in housing, transport and sustainable urban planning. Detailed modelling will clearly be of use for planning for a *post-carbon future*, one in which we stop burning fossil fuels.

For these reasons there is growing interest in spatial microsimulation. This is due largely to its practical utility in an era of 'evidence-based policy' but is also driven by changes in the wider research environment inside and outside of academia. Continued improvements in computers, software and data availability mean the methods are more accessible than ever. It is now possible to simulate the populations of small administrative areas at the individual level almost anywhere in the world. This opens new possibilities for a range of applications, not least policy evaluation.

Still, the meaning of spatial microsimulation is ambiguous for many. This book also aims to clarify what the method entails in practice. Ambiguity surrounding the term seems to arise partly because the methods are inherently complex, operating at multiple levels, and partly due to researchers themselves. Some uses of the term 'spatial microsimulation' in the academic literature are unclear as to its meaning; there is much inconsistency about what it means. Worse is work that treats spatial microsimulation as a magical black box that just 'works' without any need to describe, or more importantly make reproducible, the methods underlying the black box. This book is therefore also about demystifying spatial microsimulation.

1.3   A definition of spatial microsimulation

At this early stage, it is worth considering how spatial microsimulation has been interpreted in past work and how we define the term for the book. Depending on the available data, the aim of the research, and the interpretation of the researcher using the techniques, spatial microsimulation has two broad meanings. Spatial microsimulation can be understood either as a technique or an approach:

1. A *method* for generating spatial microdata — individuals allocated to zones (see Figure 1.1) — by combining individual and geographically aggregated datasets. In this interpretation, 'spatial microsimulation' is roughly synonymous with 'population synthesis'.
2. An *approach* to understanding multi level phenomena based on spatial microdata — simulated or real.

Figure 1.1 is a simplified flow diagram showing the difference between spatial microdata (bottom) and more commonly found official data types (top). There are clear disadvantages with both geographically aggregated and individual level (non-geographical) sources. Spatial microsimulation seeks to overcome some of these limitations, creating new outputs that are useful for a number of research applications and as inputs to more sophisticated models.

Throughout this book we use a broad definition of the term spatial microsimulation:

The creation, analysis and modelling of individual level data allocated to geographic zones.

To further clarify the distinction between the methodology and the approach, we use term *population synthesis* to describe the narrower process of generating the spatial microdata (see Chapter 5). The term *spatial microsimulation* is reserved to describe the overall process, which usually involves population synthesis. Unless you are lucky enough to have access to real spatial microdata, a scarce but increasingly available commodity, spatial microdata must be generated by combining zone level and individual level data. Throughout the course of the book the emphasis shifts: the focus in Chapters 4 to 6 is on generating spatial microdata while Chapters 7 onwards focus on how to use this rich data type.

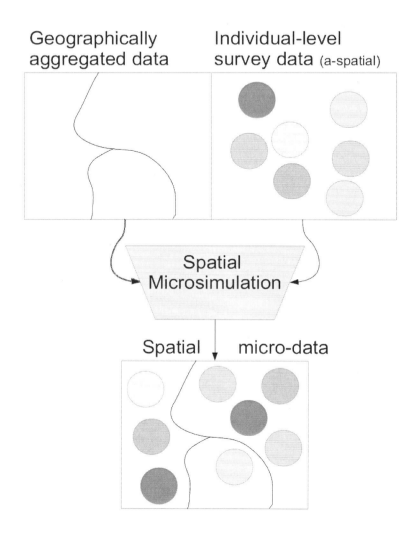

FIGURE 1.1
Schematic diagram contrasting conventional official data (above) against spatial
microdata produced during *population synthesis* (below). Note that for brevity
the geographically aggregated data are called *constraints* and the individual
level survey data are called *microdata* in this book.

1.4   Learning by doing

Another issue tackled in this book is reproducibility, which is intimately linked to the 'learning by doing' approach. Notwithstanding a few exceptions (e.g. Williamson, 2007), most findings in the field cannot be replicated: there is no way of independently verifying the results of spatial microsimulation research. Some model-based decisions, such as where to build smoking cessation clinics (Tomintz et al., 2008), are a matter of life or death. In such cases model transparency becomes vital.

Today fast internet connections, open access datasets and free software make it easier than ever to create reproducible research. This movement is growing in many fields, from Regional Science to Epidemiology (Ince et al., 2012; McNutt, 2014; Peng et al., 2006; Rey, 2014).

This book encourages 'Open Science' throughout. Reproducibility is encouraged through the provision of code. You will be expected to *do* spatial microsimulation rather than simply *watch* the results that others have produced! Small datasets are provided on which these 'code chunks' (such as that presented below) can run. All the findings presented in this book can therefore be reproduced using code and data in the book's GitHub repository (`https://github.com/Robinlovelace/spatial-microsim-book`).

```
# This is a code chunk. Below is the code!
ind <- read.csv("data/SimpleWorld/ind-full.csv") # read-in data
print(ind[1,]) # print the 1st line of the data
```

```
##   id age sex income
## 1  1  59   m   2868
```

Notice that in addition to showing what to type into R, we show what the output should be. In fact, using the 'RMarkdown' file format, the code runs every time the book is compiled, ensuring the code's long-term viability. There are good reasons to ensure such a reproducible work-flow. First, reproducibility can save time. Reproducible code that can be re-used multiple times, meaning you only have to write it once, reducing the need to 're-invent the wheel' for every new analysis. Second, reproducibility can increase research impact: by enabling and encouraging others to use your work, it can gain more credit and, if you are in academia, citations. The third reason is more profound: reproducibility is a prerequisite of falsifiability, the backbone of science (Popper, 1959). The results of non-reproducible research cannot be verified, reducing its scientific credibility. All these reasons inform the book's practical nature.

This book presents spatial microsimulation as a living, evolving set of techniques rather than a prescriptive formula for arriving at the 'right' answer. Spatial microsimulation is largely defined by its user-community, made up of a growing number of people worldwide. This book aims to contribute to the community by encouraging collaboration, innovation and rigour. It also encourages playing with the methods and 'getting your hands dirty' with the code. As Kabacoff (2011 p. xxii) put it regarding R: "The best way to learn is to experiment".

1.5   Why spatial microsimulation with R?

Software decisions have a major impact on the flexibility, efficiency and reproducibility of research. Nearly three decades ago Clarke and Holm (1987) observed that "little attention is paid to the choice of programming language used" for microsimulation. This appears to be as true now as it was then. Software is rarely discussed in papers on the subject and there are few mature spatial microsimulation packages.[1] There is a strong and diverse software community in the ABM field, and many of these bring insights that could be of use to researchers using spatial microsimulation. Mannion et al. (2012), for example, present JAMSIM. This is a new software framework for combining the powerful Repast ABM application with R for graphing and analysis of the results. JAMSIM and other projects in the field (including the development of NetLogo and MASON ABM systems and associated add-ons) should be seen as developing in parallel to the R approach developed in this field, not competitors. There is clearly much potential for mutual benefit by software for ABM and spatial microsimulation interacting more closely together.

Factors that should influence software selection include cost, maturity, features and performance. Perhaps most important for busy researchers are the ease and speed of learning, writing, adapting and communicating the analysis. R excels in each of these areas.

R is a *low level* language compared with statistical programs based on a strong graphical user interface (GUI) such as Microsoft Excel and SPSS. R offers great flexibility for analysing and modelling data and enables easy creation of user-defined functions. These are all desirable attributes of software for undertaking spatial microsimulation. On the other hand, R is *high level* compared with general purpose languages such as C and Python. Instead of writing code to perform statistical operations 'from scratch', R users generally use pre-made functions. To calculate the mean value of variable x, for example, one would

---

[1]The Flexible Modelling Framework (FMF (`https://github.com/MassAtLeeds/FMF`)) is a notable exception written in Java that can perform various modelling tasks.

need to type 20 characters in Python: `float(sum(x))/len(x)`.[2] In pure R just 7 characters are sufficient: `mean(x)`. This terseness and range of pre-made functions is useful for ease of reading and writing spatial microsimulation models and analysing the results.

The example of calculating the mean in R and Python may be trite but illustrates a wider point: R was *designed* to work with statistical data, so many functions in the default R installation (e.g. `lm()`, to create a linear regression model) perform statistical analysis 'out of the box'. In agent-based modelling, the statistical analysis of results often occupies more time than running the model itself (Thiele et al., 2014). The same applies to spatial microsimulation, making R an ideal choice due to its statistical analysis capabilities.

R has an active and growing user community and is easy to extend. R's extreme flexibility allows it to call code written in other programming languages. This means that 'glass ceilings' encountered in other environments are not an issue in R: there is a wide range of algorithms that can be used from within R. This ability has been used to dramatically speed up R code. The new R package **dplyr**, for example, uses C++ code to do the hard work of data manipulation tasks.

As a result of R's flexibility, open source ethic and strong user community, there are thousands of packages that extend R's capabilities with new functions. Improvements are being added to the 'R ecosystem' all the time. This book provides an overview of the functionality of key packages for spatial microsimulation. Software development is a fast moving game, however, so keep a look out for updates. New software can save both research and computer time so it's worth staying up-to-date with the latest developments.

The **ipfp** and **mipfp** packages, for example, can greatly reduce the number of lines of code *and* the computational time needed for population synthesis compared with 'hard-coding' the method in R from scratch. (For the curious reader, the 'IPF' letters in the package names stand for 'Iterative Proportional Fitting', one of a number of technical terms we will use frequently in subsequent chapters. See the Glossary for a definition.)

These time-saving add-ons to R are described in Chapter 5. A clear advantage of R is that anyone can write a package. This is also potentially a disadvantage: it has been argued there are too many packages, making it difficult for new users to identify which packages are most reliable and which are best suited to different situations (Hornik, 2012). However, this problem is mitigated by the open-source nature of R: users can see precisely how any particular function works (providing they are willing to learn some programming), rather than relying on a "black box".

---

[2]The `float` function is needed in case integers are used. This can be reduced to 13 characters with the excellent **NumPy** package: `import numpy; x = [1,3,9]; numpy.mean(x)` would generate the desired result. The R equivalent is `x = c(1,3,9); mean(x)`.

A recent development that has made R far more accessible is RStudio, an Integrated Development Environment (IDE) for R that helps new users become familiar with the language (see Figure 2.2). Further information about why R is a good choice for spatial microsimulation is provided in the Appendix, a tutorial introduction to R for spatial microsimulation applications. The next section describes approaches to learning R in general terms.

1.6   Learning the R language

Having learned a little about *why* R is a good tool for the job, it is worth considering at this stage *how* R should be used. It is useful to think of R not as a series of isolated commands, but as an interconnected *language*. The code is used not only for the computer to crunch numbers, but also to communicate ideas, from one person to another. In other words, this book teaches spatial microsimulation in the language of R. Of course, English is more appropriate than R for *explaining* rather than merely describing the method and the language of mathematics is ideal for describing quantitative relationships conceptually. However, because the practical components of this book are implemented in R, you will gain more from it if you are fluent in R. To this end the book aims to improve your R skills as well as your ability to perform spatial microsimulation, especially in the earlier practical chapters. Some prior knowledge of R will make reading this book easier, but R novices should be able to follow the worked examples, with reference to appropriate supplementary material.

The references listed in the Appendix (Section 13.4) provide a recommended starting point for R novices. For complete beginners, we recommend the introductory 'Owen R Guide' (Owen, 2006). As with learning Spanish or Chinese, frequent practice, persistence and experimentation will ensure deep learning.

A more practical piece of advice is to organise your workflow. Each project should have its own self-contained folder containing all that is needed to replicate the analysis, except perhaps large input datasets. This could include the raw (unchanged) input data[3], R code for analysis, the graphical outputs and files containing data outputs. To avoid clutter, it is sensible to arrange this content into folders, as illustrated below (thanks to Colin Gillespie for this tip):

```
|-- book.Rmd
```

---

[3]Raw data should be kept safely on an external hard disk or a server if it is large or sensitive.

```
|-- data
|-- figures
|-- output
|-- R
|    |-- load.R
|    `-- parallel-ipfp.R
`-- spatial-microsim-book.Rproj
```

The example directory structure above is taken from an early version of this book. It contains the document for the write-up (`book.Rmd` — this could equally be a `.docx` or `.tex` file) and RStudio's `.Rproj` file in the *source directory*. The rest of the entities are folders: one for the input data, one for figures generated, one for data outputs and one for R scripts. The R scripts should have meaningful names and contain only code that works and is commented. An additional backup directory could be used to store experimental code. There is no need to be prescriptive in following this structure, but projects using spatial microdata tend to be complex, so imposing order over your workflow early will likely yield dividends in the long run. If you work in a team, this kind of structure helps each member to rapidly understand and continue your work.

The same applies to learning the R language. Fluency allows complex numerical ideas to be described with a few keystrokes. If you are a new R user it is therefore worth spending some time learning the R language. To this end the Appendix provides a primer on R from the perspective of spatial microsimulation.

1.7 Typographic conventions

The following typographic conventions are followed to make the practical examples easier to follow:

- In-line code is provided in `monospace` font to show it's something the computer understands.
- Larger blocks of code, referred to as *listings*, are provided on separate lines and have coloured *syntax highlighting* to distinguish between values, names and functions:

```r
x <- c(1, 2, 5, 10) # create a vector
sqrt(x) # find the square root of x
```

```
## [1] 1.000000 1.414214 2.236068 3.162278
```

- Output from the *R console* is preceded by the ## symbol, as illustrated above.
- Comments are preceded by a single # symbol to explain specific lines.
- Often, reference will be made to files contained within the book's project folder. The notation used to refer to the location of these files follows the way we refer to files and folders on Linux computers. Thus 'R/CakeMap.R' refers to a file titled 'CakeMap.R', within the 'R' directory of the project's folder.

There are many ways to write R code that will generate the same results. However, to ensure clarity and consistency, a single style, advocated in a chapter (http://r-pkgs.had.co.nz/style.html) in Hadley Wickham's *Advanced R*, is followed throughout (Wickham, 2014a). Consistent style and plentiful comments will make your code readable by yourself and others for decades to come.

1.8 An overview of the book

The book's chapters progress in a logical order that corresponds to the steps typically taken during a spatial microsimulation research project that generates, analyses and models spatial microdata. The next two chapters are primarily conceptual, introducing the concept and applications of spatial microsimulation in more detail whilst the remaining chapters are practical. Because we'll be making heavy use of terminology specific to spatial microsimulation, it is recommended that you read these introductory chapters before tackling the practical section that follow although this is not essential. If any of the words do not make sense, the Glossary at the end of the book may help to clarify what they mean.

In Part II, Chapters 4, 5, 6 and 7 cover in detail the process of *population synthesis* whereby spatial microdata is generated. This process is key to much spatial microsimulation work. These chapters should ideally be read together in that order, although Chapter 5 is the most important and provides the basics needed to create spatial microdata. Chapters 8 introduces the important step of model validation.

In Part III, Chapters 9 and 10 cover more advanced methods for generating spatial microdata when no individual level data is available and when household level data are required. These chapters are more self-standing and can be read in any order. The chapter titles and brief descriptions of their contents are as follows:

- *SimpleWorld* (Chapter 2) is a simple and reproducible explanation of spatial microsimulation with reference to an imaginary planet. It's also a chance to set up your system and RStudio settings to get the most out of the subsequent chapters.

- *What is spatial microsimulation?* (Chapter 3) introduces the concepts and applications of spatial microsimulation not only as a narrow methodology but as an *approach* to understanding real-world phenomena.

- *Data preparation* (Chapter 4) is dedicated to the boring but vital task of loading and 'cleaning' the input data, ready for spatial microsimulation.

- *Population synthesis* (Chapter 5) uses the input data created in the previous chapter to perform population synthesis, the process of creating synthetic spatial microdata. The chapter describes the main functions and techniques that are used for undertaking this key aspect of spatial microsimulation, with a focus on the **ipfp** and **mipfp** packages, which perform Iterative Proportion Fitting (IPF).

- *Alternative approaches to population synthesis* (Chapter 6) describes some alternative methods, beyond IPF, for population synthesis. These include an R implementation the Generalized Regression Weighting procedure (GREGWT), a description of 'simulated annealing' and a demonstration that population synthesis can be interpreted as a constrained optimisation problem. This final insight has the consequence that many previously developed optimisation algorithms can be used to perform the 'heavy lifting' of population synthesis. A few of these, including R's `optim` and `GenSA` functions, are tested, illustrating the importance of performance testing.

- *Spatial microsimulation in the wild* (Chapter 7) is based on a larger and more complex example than SimpleWorld called CakeMap, that uses real data. This chapter combines geographic constraints with non-geographical survey data to estimate the geographical variation in cake consumption.

- *Model checking and evaluation* (Chapter 8) tackles the understudied but important questions of 'is the model working?' and 'do the results coincide with reality?'. Here you will learn appropriate tests and checks to verify spatial microsimulation outputs and think about its underlying assumptions.

- *Population synthesis without microdata* (Chapter 9) gives a way to produce a spatial microsimulation without having individual level data.

- *Household allocation* (Chapter 10) explains techniques to group the synthetic individuals into households.

- *The TRESIS approach to spatial microsimulation* (Chapter 11) describes an alternative approach to spatial microsimulation that operates at the household level. This contributed chapter, by Richard Ellison and David

Hensher, demonstrates the population synthesis stage of the wider Transport and Environment Strategy Impact Simulator (TRESIS) modelling system.

- *Spatial microdata in agent-based models* (Chapter 12) describes how the output from spatial microsimulation can be used as an input into more complex models.

The final thing to say in this opening chapter relates to the future of this book as an up-to-date and accessible teaching resource. *Spatial Microsimulation with R* is both a book and an open source, open access project that is designed to be improved on and updated by the authors and the community using spatial microsimulation. Any comments relating to code, content or clarity of explanation will therefore be gratefully received.[4]

---

[4]Feedback can be left via email to r.lovelace@leeds.ac.uk (`mailto:r.lovelace@leeds.ac.uk`) or via the project's GitHub page (`http://github.com/Robinlovelace/spatial-microsim-book`).

2

SimpleWorld: A worked example of spatial microsimulation

CONTENTS

This chapter focuses on the minimum input datasets needed for the classical type of microsimulation. We will use input data on the inhabitants of an imaginary world (geographical individual level) called SimpleWorld to demonstrate the basic concepts and techniques to perform spatial microsimulation with R.

This is the first practical chapter and it aims to acquaint you with language R and the R development program RStudio. Fluency with R and RStudio, in combination with well-organised project management and clear, commented code, will enable fast and efficient work. In short, this chapter aims to teach good 'workflow' in preparation for navigating the subsequent chapters. A secondary aim is to illustrate key concepts in spatial microsimulation with reference to a practical example, called 'SimpleWorld' for its simplicity.

The chapter also serves to highlight the links between the methodology presented in-depth later in the book (see Chapter 5) and the various real-world applications presented in Chapter 4. For R 'newbies' it also provides a chance for the reader to do some basic programming in R: the unwritten subtitle of this book is "a *practical* introduction" for a reason!

This chapter is ordered as follows:

- *Getting setup with the RStudio environment* (Section 2.1) contains the basic explanation of RStudio to enable you to code the examples of this book on your own computer.
- *SimpleWorld data* (Section 2.2) describes the context and the data of this little illustrative example.
- *Generating a weight matrix* (Section 2.3) contains the code and the description of the role of a weight matrix in a spatial microsimulation. *-Spatial microdata* (Section 2.4) shows the typically useful output of a spatial microsimulation. *-SimpleWorld in context* (Section 2.5) mentions the kind of context where the output could be used.

2.1 Getting setup with the RStudio environment

Before progressing further, it is important to ensure that R is setup correctly and working on your computer, so you can follow the practical examples. This section will not go into much detail. Further resources are provided in the Appendix at the end of the book to provide insight into how R works as a programming language.

The majority of the help that you need to install and setup R on your computer can be found online, so it is recommended that you consult this section with reference to online searches and help pages. As with all open source software projects R evolves over time so it is important to keep up-to-date with any changes, which may not be accounted for in this book.

2.1.1 Installing R

The practical examples in this book require a recent version of R to be installed on your computer. This will allow you to type test the code as you learn the methods. To install R, we recommend you refer to documentation provided online, on the Comprehensive R Archive Network (CRAN):

- For Windows, see `https://cran.r-project.org/bin/windows/base/`
- For Mac, see `https://cran.r-project.org/bin/macosx/`
- For Linux, see `https://cran.r-project.org/doc/manuals/r-release/R-admin.html`

On Debian-based systems such as Ubuntu, the following bash command should be sufficient to install R.

```
sudo apt-get install r-base
```

Once R is installed, new packages are easy to install from within R using `install.packages("package_name")`. To install the **mipfp** package, for example, we would use:

```
install.packages("mipfp")
```

2.1.2 RStudio

RStudio is an Integrated Development Environment (IDE) for R. It provides an advance Graphical User Interface (GUI) that makes it easier not only to write R code, but to install packages, manage R's graphical outputs and organise complex projects. All R code presented in this book will run in other environments, such as Linux's bash shell, in which R can be initiated by the following command:

```
$ R # enter the R command line
```

It is highly recommended that you use RStudio for the work presented in this book, however, especially if you an R beginner. RStudio will enable you to spend less time worrying about how to write R code correctly and more time focussed on spatial microsimulation. One example of how RStudio saves time is its tab autocompletion functionality. To test this functionality, type the beginning of any R object or function stored in R's environment (such as `plo`, short for `plot`) and hit the `Tab` button. The full name of various options should auto-complete, as illustrated in Figure 2.1. Pressing the down arrow allows the user to select the correct input. Auto-completion allows fast typing, easy recollection of function/object names. It also prevents errors cause by typos in your code.

To test this functionality, download and install RStudio if you have not already. The installation process is simple on Linux, Mac and Windows platforms: simply follow the instructions provided here at rstudio.com/products/rstudio/download/.

There are many other advantages of using RStudio.[1] The next section focuses on one particular advantage that we recommend you make use of whilst working through the book: RStudio projects.

---

[1]For more information about RStudio, please see the RStudio website: `https://www.rstudio.com/products/rstudio/features/` and search for other on-line resources. There is even an entire book dedicated to the subject (Verzani, 2011).

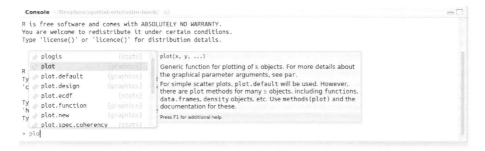

FIGURE 2.1
Autocompletion in RStudio

2.1.3   Projects

RStudio provides a neat system for creating, managing and even sharing your projects. This functionality merits a section of its own, because spatial microsimulation projects are likely to be large and complex, requiring collaboration with other people. If you are not careful the projects can become over-complex and unmanageable. RStudio projects can prevent this from happening and lead to improved workflow.

To setup a new project click on the drop-down menu in the top-right section of RStudio. If you are working on a project, the name of the project will appear in this dropdown menu (see Figure 2.2). When a project is loaded, the following things happen:

- The files open in the script panel (top left) last time you worked on the project will appear.
- R's working directory (set and checked with `setwd()` and `getwd()`, respectively) will change.
- Any R objects saved in the `.RData` file in the project will be automatically loaded, saving your work.
- If you are using `git` to manage your project's development and sharing, you'll have options to commit code and 'push' new work, e.g. to GitHub.

In fact, this entire book was written as an RStudio project. This allowed us to work from the same system and share code easily, by pushing our code frequently to the book's online repository: `https://github.com/Robinlovelace/spatial-microsim-book` . This online repository will allow you to access all of the example code and data for the book. It will also allow you to contribute to the book as an evolving project.

The next section explains how to download and use the data stored in the `spatial-microsim-book` GitHub repository to access the code and data resources associated with the book.

2.1.4 Downloading data for the book

To simply download the code and data, associated with this book, first navigate to `https://github.com/Robinlovelace/spatial-microsim-book` in your browser. From this page, click on the 'Download ZIP' button to the right and extract the folder into a sensible place on your computer, such as the Desktop.[2]

With the files safely stored on the computer, the next step is to open the folder as an RStudio project. To do this, open the folder in a file browser and double click on `spatial-microsim-book.Rproj`. Alternatively, start RStudio and open the file by clicking on 'Open Project' from the dropdown menu above the top-right panel, or by clicking with File > Open. This will cause RStudio to open the project. All the input data files should now be easily accessible to R through *relative file paths*. The files should also be visible in the Files tab in bottom right panel.

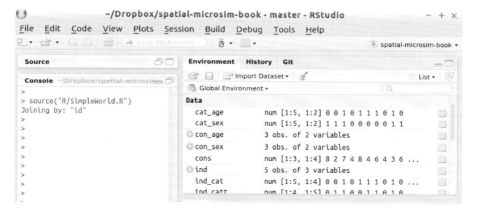

FIGURE 2.2
The RStudio interface with the spatial-microsim-book project loaded and objects loaded in the Environment after running SimpleWorld.R

To check if the project has been downloaded and opened correctly, try running the following code, by typing it and hitting `Ctl-Enter`:

---

[2]An alternative way to get the project files onto your computer is to 'clone' the repository. For further details on how to fork, clone and potentially contribute to the book project, see the GitHub website, or refer to the growing literature how GitHub can be used in research (Lima et al., 2014).

```
source("R/SimpleWorld.R")
```

You should see some output in red, beginning `Attaching package: 'dplyr'`. Some new objects should also appear in the Environment tab in the top-right panel (Figure 2.2). This is the result of running the code located in a script file called `SimpleWorld.R`, using the `source()` function.

Now that you have an understanding of RStudio and how to load the book project, it's time to 'get your hands dirty' and run some code! In the subsequent section we will type and run some of the contents of the `SimpleWorld.R` script line by line.

2.2 SimpleWorld data

SimpleWorld is a small world, consisting of 33 persons split across 3 zones, as illustrated in Figure 2.3. We have two sources of information about these people available to us:

1. aggregate counts of persons by age and sex in each zone (from the SimpleWorld Census)
2. survey microdata recording more detailed information (age, sex and income), for five of the world's residents.

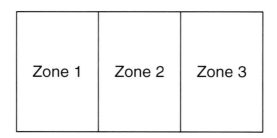

FIGURE 2.3
The SimpleWorld environment, consisting of individuals living in 3 zones.

Unfortunately the survey data lack geographical information, and include only a small subset of the population (5 out of 33). To infer further information about SimpleWorld — and more importantly to be able to model its inhabitants — we need a methodology. This is precisely the kind of situation where spatial microsimulation is useful. After an explanation of this 'starting simple' approach, we describe the input data in detail, demonstrate what spatial microdata look like and, finally, place the example in its wider context.

This SimpleWorld example is analogous to the 'cartoon world' used by professor MacKay (2008) to simplify the complexities of sustainable energy. The same principle works here: we will generate a synthetic population for the imaginary SimpleWorld to illustrate how spatial microsimulation works and how it can be useful.

The SimpleWorld planet, a 2 dimensional plane split into 3 zones, is illustrated in Figure 2.3. SimpleWorld is inhabited by 12, 10 and 11 individuals of its alien inhabitants in zones 1 to 3, respectively: a planetary population of 33. From the SimpleWorld Census, we know how many young (strictly under 50 space years old) and old (50 and over) residents live in each zone, as well their genders: male and female. This information is displayed in the tables below.

zone	0-49 yrs	50 + yrs
1	8	4
2	2	8
3	7	4

TABLE 2.1: Aggregate level age counts for SimpleWorld.

Zone	m	f
1	6	6
2	4	6
3	3	8

TABLE 2.2: Aggregate sex counts for SimpleWorld.

The following commands load-in this data.

```
# Load the constraint data
con_age <- read.csv("data/SimpleWorld/age.csv")
con_sex <- read.csv("data/SimpleWorld/sex.csv")
cons <- cbind(con_age, con_sex)
```

We recommend R beginners interact with the R objects just loaded: try printing them to screen, plotting them and even-subsetting them. If you cannot, it may be worth consulting the Appendix on using R, or referring to R's copious online documentation. The data contained in the `cons` object presented below refer to the characteristics of inhabitants in different SimpleWorld zones.

```
cons # print the constraint data to screen
```

```
##    a0.49 a.50. m f
## 1      8     4 6 6
## 2      2     8 4 6
## 3      7     4 3 8
```

Each row represents a new zone (see also Tables 2.1 and 2.1). The planet's entire population is represented in the counts in these tables. From these *constraint tables* we know the marginal distributions of the two categorical variables, age and sex. The data does not tell us about the contingency table (or cross-tabulation) that links age and sex. This means that we have per-zone information on the number of young and old people and the number of males and females. But we currently lack information about the number of young females, young males, and so on. Also note that we have no information about other important variables such as income.

Spatial microsimulation can be used to better understand the population of SimpleWorld. To do this, we need some additional information: an individual level *microdataset*.

This is provided in an individuals level dataset on 5 of SimpleWorld's inhabitants (Table 2.3). Note that the individual level data has different dimensions than the aggregate data presented above.

id	age	sex	income
1	59	m	2868
2	54	m	2474
3	35	m	2231
4	73	f	3152
5	49	f	2473

TABLE 2.3: Individual level survey data from SimpleWorld.

The individual level *microdata* survey has one row per individual whereas the *constraints* have one row per zone. This individual level data includes two variables that *link* it to the aggregate level data described above: age and sex.

The individual level data also provides information about a *target variable* not included in the geographical constraints: income. To load the individual level data, enter the following:

```
ind <- read.csv("data/SimpleWorld/ind-full.csv")
```

Note that although the microdataset contains additional information about the inhabitants of SimpleWorld, it lacks geographical information about where each inhabitant lives or even which zone they are from. This is typical of individual level survey data. Spatial microsimulation tackles this issue by allocating individuals from a non-geographical dataset to geographical zones in another.

2.3 Generating a weight matrix

The process to generate spatial microdata is allocating *weights* to each individual for each zone. The higher the weight for a particular individual-zone combination, the more representative the individual is of that zone. This information can be represented as a *weight matrix*.

The subsequent code block uses the **mipfp** package to convert these inputs into a matrix of weights. Just type it in and observe the result. If it works, congratulations! You have generated your first weight matrix using Iterative Proportional Fitting (IPF) (see code output). The details of this process are described in subsequent chapters.

```
target <- list(as.matrix(con_age[1,]), as.matrix(con_sex[1,]))
descript <- list(1, 2)
ind$age_cat <- cut(ind$age, breaks = c(0, 50, 100))
seed <- table(ind[c("age_cat", "sex")])

library(mipfp) # install.packages("mipfp") is a prerequisite
res <- Ipfp(seed, descript, target)
res$x.hat # the result for the first zone
```

```
##              sex
## age_cat            f        m
##    (0,50]   4.455996 3.544004
##    (50,100] 1.544004 2.455996
```

Note that the output of the above command is only for the first zone. It suggests that in zone 1, younger females are far more (nearly 4.5 times more) numerous than would be expected based on the individual level data alone. This makes sense because `con_age[1,]` contains many young people and `ind` contains only one young female. At this stage you may be wondering: what just happened? This is explained in detail in the subsequent chapters.

An alternative approach to arriving at a similar result, for all zones, is demonstrated in the code chunk below using another package: **ipfp**. The result of this code is illustated in Table 2.4. Again, the reader is not expected to understand what has just happened. This will be explained in subsequent chapters.

```
A <- t(cbind(model.matrix(~ ind$age_cat - 1),
         model.matrix(~ ind$sex - 1)[, c(2, 1)]))

cons <- apply(cons, 2, as.numeric) # convert to numeric data

library(ipfp) # install.packages("ipfp") is a prerequisite
weights <- apply(cons, 1, function(x) ipfp(x, A, x0 = rep(1, 5)))
weights[,1] # result for the first zone
```

```
## [1] 1.227998 1.227998 3.544004 1.544004 4.455996
```

Individual	Zone 1	Zone 2	Zone 3
1	1.23	1.73	0.725
2	1.23	1.73	0.725
3	3.54	0.55	1.550
4	1.54	4.55	2.550
5	4.46	1.45	5.450

TABLE 2.4: A 'weight matrix' linking the microdata (rows) to the zones (columns) of SimpleWorld.

The highest value (5.450) is located, to use R's notation, in cell `weights[5,3]`, the 5th row and 3rd column in the matrix `weights`. This means that individual number 5 is considered to be highly representative of Zone 3, given the input data in SimpleWorld. This makes sense because there are many (7) young people and many (8) females in Zone 3, relative to the input microdataset (which contains only 1 young female). The lowest value (0.55) is found in cell `[3,2]`. Again this makes sense: individual 3 from the microdataset is a young male yet there are only 2 young people and 4 males in zone 2. A special feature of the weight matrix above is that each of the column sums is equal to the total population in each zone.

Note that the first method gives weights to categories (for example young female), whereas the second method gives weights to each individual. This is normal and equivalent, we can easily transform one type of output in the other type. The details are explained further.

We will discover different ways of generating such weight matrices in subsequent chapters. For now it is sufficient to know that the matrices link individual level data to geographically aggregated data and that there are multiple techniques to generate them. The techniques are sometimes referred to as *reweighting algorithms* in the literature (Tanton et al. 2011). These include deterministic methods such as Iterative Proportional Fitting (IPF) and probabilistic methods that rely on a pseudo-random number generator such as simulated annealing (SA). These and other reweighting algorithms are described in detail in Chapter 5.

2.4 Spatial microdata

A useful output from spatial microsimulation is what we refer to as *spatial microdata*. This is a dataset that contains a single row per individual (as with the input microdata) but also an additional variable on geographical location.

The ideal spatial microdataset selects individuals representative of the aggregate constraints for each zone, while containing the diversity of information present in the individual level non-spatial input population. In Chapter 4 we will explore all the steps needed to produce a spatial microdataset for SimpleWorld. A subset of such spatial microdataset is presented in Table 2.5, where each row represents an individual taken from the individual level sample and the 'zone' column represents the geographical zone in which they reside.

id	zone	age	sex	income
1	2	59	m	2868
2	2	54	m	2474
4	2	73	f	3152
4	2	73	f	3152
4	2	73	f	3152
4	2	73	f	3152

id	zone	age	sex	income
5	2	49	f	2473
4	2	73	f	3152
5	2	49	f	2473
2	2	54	m	2474

TABLE 2.5: Spatial microdata generated for SimpleWorld zone 2.

Table 2.5 is a reasonable approximation of the inhabitants of zone 2: older females dominate in both the aggregate (which contains 8 older people and 6 females) and the simulated spatial microdata (which contains 8 older people and 6 females). Note that in addition to the constraint variables, we also have an estimate of the income distribution in SimpleWorld's second zone.

By the end of Chapter 5, you should learn how to generate this table and understand each of the steps involved. The remainder of this section considers how the outputs of spatial microsimulation, in the context of SimpleWorld, can be useful before progressing to the practicalities.

2.5   SimpleWorld in context

Even though the datasets are tiny in SimpleWorld, we have already generated some useful output. We can estimate, for example, the average income in each zone. Furthermore, we could create an estimate of the *distribution* of income in each area. Although these estimates are unlikely to be very accurate due to the paucity of data, the methods could be very useful if performed on larger datasets from 'RealWorld' (planet Earth). The spatial microdata presented in the above table could be used as an input into an agent-based model (ABM). Assuming the inhabitants of SimpleWorld are more predictable than those of RealWorld, the outputs from such a model could be very useful indeed, for example for predicting future outcomes of current patterns of behaviour.

In addition to clarifying the advantages of spatial microsimulation, the above example also flags some limitations of the methodology. Spatial microsimulation will only yield useful results if the input datasets are representative of the population as a whole, and for each region. If the relationship between age and sex is markedly different in one zone compared with what we assume to be the global averages of the input data, for example, our estimates could be way off Using such a small sample, one could rightly argue, how could the diversity of 33 inhabitants of SimpleWorld be represented by our simulated spatial microdata? This question is equally applicable to larger simulations. These issues are important and will be tackled in Chapter 8.

2.6   Chapter summary

In this chapter we have learned how to setup the R/RStudio environment for running the example code provided in this book. This involved installing the software, understanding projects in RStudio and setting up the `spatial-microsim-book` project so that data can be accessed easily. This was the first practical chapter and it involved basic commands for loading data and creating a weight matrix. The process of running and playing with the code should have helped get acquainted with the 'workflow' associated with developing and running spatial microsimulation models in R. Discussion of the results, which represent the inhabitants of the imaginary system *Simple-World*, create the backdrop for the rest of the book, which we will return to in Chapter 12.

3

What is spatial microsimulation?

CONTENTS

The purpose of this chapter is to demystify the various interpretations and uses of the term 'spatial microsimulation' and to define clearly what we mean by it in this book. Following the brief introduction to the field in this first section, the chapter is ordered as follows:

- *Terminology* (Section 3.1) explains the basis of the concept (Section 3.1.1) and shows how spatial microsimulation can be understood as either a method or an approach (Section 3.1.2).
- *What spatial microsimulation is not* (Section 3.2) describes the differences between spatial microsimulation and other fields that could seem similar.
- *Applications* (Section 3.3) explores various academic and real-world applications.
- *Assumptions* (Section 3.4) addresses the often unspoken assumptions underlying spatial microsimulation results.

For the purposes of this book spatial microsimulation was defined in Chapter 1 as:

The creation, analysis and modelling of individual level data allocated to geographic zones.

Spatial microsimulation is well-suited to the analysis of complex phenomena which happen over geographical space, such as transport systems and housing markets. Because it includes the creation of synthetic data, the method is well-suited to situations in which available data are limited. Figure 3.1 illustrates how the process of *population synthesis* can be used to impute missing data, by approximating the original geo-referenced individual level data (Lovelace et al., 2015). It is important to note that the process does not stop with the generation of spatial microdata: it involves *doing stuff* with the spatial microdata to better understand the world.

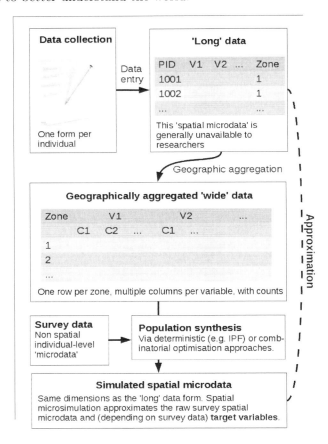

FIGURE 3.1
Schematic of population synthesis, a critical element in spatial microsimulation

Spatial microsimulation is a way to combine the advantages of individual level data with the geographical specificity of geographical data. If used correctly, it can be used to provide 'the best of both worlds' of available data by combining them into a single analysis. Spatial microsimulation should therefore be used in situations where you have access to individual and area level data but no means of combining them. Spatial microsimulation can also be used in contexts where no individual level data is available, as described in Chapter 9.

Typical use cases include modelling the spatial distribution of population growth and changes in demographics; high level transport-modelling; scenario-based modelling of behavioural change; and as an input into agent-based models. Because large individual and area level datasets have only recently become available and because computers have only recently become sufficiently powerful to run large-scale models on thousands of individuals and zones, spatial microsimulation is still a field in its infancy. There are many areas of application where the method has considerable benefits but where the method has not yet been used. It is hoped that this book provides sufficient guidance to enable the reader to use spatial microsimulation on new problems.

3.1 Terminology

Like many phrases, the meaning of *spatial microsimulation* varies depending on context. The term is more ambiguous than most, however, because it is a relatively new method and because different people have used the term in different ways. Of course, the label attributed to the method is less important than the method itself. As eminent physicist Richard Feynman put it, "the difference between knowing the name of something and knowing something" is vital to understand the world. The methods used in this book could equally be "multi level modelling" or "real-world SimCity", but this would not change how its methods work or what they do. However, it is important that the terminology we use is at least *internally consistent* to avoid confusion. Furthermore, it is important to understanding something about how others have used 'spatial microsimulation' to avoid misinterpretation of literature that employs the term.

There are some issues with term *spatial microsimulation*, as described below (see Section 3.2). However, *spatial microsimulation* is an appropriate label for the material covered in this book because it is already widely used and because it succinctly conveys the main elements of the approach:

- **Spatial** microsimulation is inherently concerned with how things vary over space, not just between individuals, groups or periods of time: this is what distinguishes *spatial* microsimulation from the wider field of microsimulation.
- Spatial **micro**simulation explores issues at the individual level, as implied by the word *micro*.

- Spatial micro**simulation** involves the creation of fictitious data for modelling purposes, captured by the word *simulation*.

From this breakdown of spatial microsimulation into its component parts its meaning may seem obvious. However, it is important to define the term precisely at an early stage and understand how other people have used the term to avoid confusion.

3.1.1 Spatial microsimulation as SimCity

SimCity is a popular computer game series in which the player constructs urban infrastructure and observes his God-like influence on the virtual citizens. The analogy of SimCity helps to describe spatial microsimulation. In practice, however, the underlying aims of SimCity (entertainment, education and profit for its publisher) are quite different from those of spatial microsimulation research. But, in some ways the comparison is appropriate: SimCity creates virtual individuals allocated to geographic space and provides a framework for *model experiments* in the same way that spatial microsimulation does. SimCity can be used for teaching urban planning (Gaber, 2007) and illustrates how complex computer simulations of urban systems can become. A number of open-source versions of the SimCity concept are now available (e.g. LinCity-NG, Micropolis and Simutrans).

3.1.2 Spatial microsimulation: method or approach?

The most common confusion about what spatial microsimulation is arises because the term has been used to refer to both a narrow methodology and a broader research approach. From Feynman's distinction between a thing's label and what it actually is, it is clear that both interpretations are acceptable. The critical step is to use the term in a way that others can understand, remembering that the audience may not have heard of spatial microsimulation before. As stated in the introduction, we acknowledge both uses of the term in the literature but advocate that 'spatial microsimulation' is used to describe the overall modelling approach. The term *population synthesis*, used in transport modelling, is used to describe the methods for generating the spatial microdata on which the spatial microsimulation approach depends.

A subsequent section (Section 3.3) outlines some real-world problems spatial microsimulation has been used to tackle. First, we consider how spatial microsimulation is understood in the literature and how this relates to the definitions used in this book.

Population synthesis is a set of techniques for generating individual level data allocated to geographical zones. Population synthesis is an important (and often crucial) component of the spatial microsimulation approach, the aim of which is to generate a realistic sample for each area that is as similar as possible to aggregate level constraints. Population synthesis usually involves the allocation of individuals from a survey dataset to administrative zones, based on shared variables between the areal (where each unit is a zone) and individual level data. When additional *target variables* exist in the microdata inputs for population synthesis (which are not present in the aggregate level data), the process can be used to simulate information that is not otherwise available at the local level. Population synthesis in this context can be seen as part of the long-established field of *small area estimation* (Rao, 2003).

Microsimulation (of which spatial microsimulation is a subset) is an approach that was first conceived by Guy Orcutt. This can be defined in general terms as "a methodology ... to simulate the states and behaviours of different units — e.g. individuals, households, firms — as they evolve in a given environment" (Baroni and Richiardi, 2007). The defining feature of *spatial microsimulation* is that the 'environment' is defined in predominantly geographical terms: the individuals are allocated to small parcels of land which affect their characteristics and inferred behaviour. This wider perspective helps explain why, despite not using the term 'spatial microsimulation', Orcutt (1957) is frequently cited as one of the founding fathers of the field.

As with many new and infrequently used words, the term spatial microsimulation is a source of confusion, and its meaning can vary depending on context and who you ask. To an economist, spatial microsimulation is likely to imply modelling a temporal process such as how individual agents in different areas respond to changes in prices or policies. To a transport planner, the term may imply simulating the precise movements of vehicles on the transport network. To your next door neighbour it may mean you have started speaking gobbledygook! Hence the need to consider what spatial microsimulation is, and what it is not, at the outset. However, in every case, the term involves the creation of individual level data that is grouped by geographic zone via some kind of approximation method.

To avoid confusion regarding the terminology used in this book, a glossary defining much of the jargon relating to spatial microsimulation is provided at the end. For now, to help answer what spatial microsimulation is we will look at its applications and then at what it is not.

3.2   What spatial microsimulation is not

Having seen contemporary definitions of spatial microsimulation and what it *is*, it is also useful to define spatial microsimulation negatively, in terms of what it is not. This is partly due to the close association between spatial microsimulation and other methods, but also because there is a tendency for people to think that spatial microsimulation is more complicated than it is.

Spatial microsimulation is not small area estimation

Small area estimation consists in estimating aggregate counts for a small area. For example, in this field, we can forecast the total population of a zone for a future year. However, we have no information about each individual, it is restricted to statistics on the area. On the other hand, spatial microsimulation really focuses on the *micro* level. Thus, we estimate the population individual per individual.

Note that, thanks to spatial microsimulation, we are able to aggregate counts and deduce the *macro* level per municipality, which corresponds to small area estimation.

Spatial microsimulation is not (quite) agent-based modelling

Spatial microsimulation does involve the creation and analysis of individuals and their allocation to families and zone. But it does not necessarily imply interaction between these individuals. For this, agent-based model (ABM) is needed. One could assume that because the method contains the word 'simulation', it includes detailed modelling of individual behaviours in which individuals interact with each other and the environment over time and space. This is not always the case: spatial microsimulation is generally a more 'top down' approach to modelling, in which the results can be broadly predicted. ABM, by contrast, is bottom-up and can result in highly non-linear and chaotic states (Batty, 2005).

There is, however, no clearly defined boundary stating where spatial microsimulation ends and ABM begins and the two approaches are closely linked. The synthetic populations produced as part of spatial microsimulation can form an excellent starting point for ABM. ABM can be seen as an extension of spatial microsimulation. While spatial microsimulation *produces* individuals and assigns their characteristics over space (and en masse via various 'what-if' scenarios), ABMs tend to have higher spatial and temporal resolution, allowing individuals to interact through space and time, with each other and with their environment. This increased level of detail and complexity means that ABM tends to have higher computational needs per individual. As a result, spatial microsimulation models tend to be much larger, encapsulating up to millions of individuals. As computing power continues to increase the potential for adding ABM capabilities to such models is only set to grow (Wegener, 2011).

The above discussion illustrates that microsimulation can be seen as a subset of advanced ABM. The results of spatial microsimulation models tend to apply to only specific snapshots in time and individuals tend to be fixed to a single area. ABM can thus add great additional value to spatial microsimulation models, by providing a framework for more complex interactions. Chapter 12 illustrates how the outputs of spatial microsimulation can form an empirical basis for ABM.

In summary, ABM and spatial microsimulation are closely related, overlapping and complementary approaches to the analysis of individual level processes operating over geographical space and time. A main conceptual difference is that the spans of space and time tend to be larger in spatial microsimulation work, for reasons relating to computing power and model complexity.

Spatial microsimulation does not really generate new data

During spatial microsimulation, apparently new individuals are created and placed into zones. It would be tempting to think that new information about the world is somehow being created. This is not the case: the 'new' individuals are simply repeats of individuals we already knew about from the individual level data, albeit in a different order and in different combinations. Thus we are not increasing the diversity of the dataset, simply changing its aggregate level characteristics. Spatial microsimulation creates a complete data that take into accounts all other data you included in the process. It is just a way to put all information together to approximate the whole real population, but it does not really generate new data.

Spatial microsimulation is often not strictly spatial

The most surprising feature of spatial microsimulation is that the method is not strictly *spatial*. The only reason why the method has developed this name (as opposed to 'small area population synthesis', for example) is that practitioners tend to use administrative zones, which represent geographical areas, as the grouping variable. However, any mutually exclusive grouping variable, such as age band or number of bedrooms in your house, could be used. Likewise, geographical location can be used as a *constraint variable*. In most spatial microsimulation models, the spatial variable is a mutually exclusive grouping, interchangeable with any such group. Of course, the spatial microdata, maps and analysis that result from spatial microsimulation are spatial; it's just that there is nothing inherently spatial about the method used to generate the spatial microdata.

To be more precise, spatial microsimulation is not *inherently spatial*. Spatial attributes such as the geographic coordinates of the zones to which individuals have been allocated and home and work locations can easily be added to the spatial microdata after they have been generated. It is the use of geographical variables as the grouping variable that is critical here and which distinguishes spatial microsimulation from other types of microsimulation.

A common use of spatial microsimulation (at least the population synthesis aspect) is simply to create model estimates of data which does not exist. This usage case is represented in Figure 3.1, whereby individual level data from a survey is 'scaled down' to the local level using population synthesis algorithms. As illustrated in Figure 3.1, the process of population synthesis can be seen as an attempt to reproduce the real spatial microdata collected during the census but which is unavailable for confidentiality reasons.

Input *microdata* and *constraints* ensure the simulated results match reality (at least at the aggregate level for the constraint variables — see Section 3.4). The resulting synthetic spatial microdata is extremely useful for estimating missing data at the local level. If *target variables* contained in the output were not present in the constraints (income is a common example), estimates of income variability over space can be extracted from the spatial microdata. In addition, the estimated spatial microdata represented in Figure 3.1 will contain estimates of *cross-tabulations* (contingency tables) between different variables and estimates of the distribution of continuous variables such as age and income. These estimates are useful in many applications (Section 3.3).

3.3 Applications

Spatial microsimulation has a wide variety of applications and there are many areas where the technique has been used. The main areas of application have been health, economic policy evaluation and transport. Rather than attempt to provide a comprehensive account of the range of current and possible applications, this section describes a single study in each area to exemplify how spatial microsimulation is used.

3.3.1 Health applications

A classic example of the potential practical utility of spatial microsimulation is a study which estimated the rate of smoking at the small area level in the city of Leeds UK (Tomintz et al., 2008). Smoking is a classic 'target variable' in spatial microsimulation: it is reported in a number of individual level surveys but there is surprisingly little information about how smoking rates vary from place to place. Thus it is difficult to determine where to locate services that depend on the rate of smoking. The synthetic spatial microdata could thus be used to help identify new clinics to help people stop smoking. (Alternatively, the spatial microdata could be used by a tobacco chain to help decide where to invest in a new shop, highlighting the potential misuse of the technique by unscrupulous analysts.) The authors found that actual anti-smoking clinics were not located optimally. Furthermore, the results pointed to optimal locations for new clinics, potentially improving the cost-effectiveness of public health campaigns.

This research has since been 'scaled-up' to estimate smoking rates across the whole of Austria. The simSALUD (`http://www.simsalud.org/`) portal provides users with access to the resulting spatial microdata and an on-line interface to allow for the selection of constraint variables and other options to customise the model for the specific purposes. This portal-based system and the provision of synthetic spatial microdata to researchers illustrates one possible direction that spatial microsimulation research could go in, where the synthetic data produced from a large model is the main output of the research, to be used by others for a variety of applications.

The example of smoking demonstrates the increase of spatial resolution that spatial microsimulation can bring to bear on under-studied areas in public health. Where the prevalence of unhealthy activities is closely related to socio-demographic variables, a synthetic microdataset can lead to decision making tools that would be difficult to implement with non-spatial surveys alone. **Simobesity** is another research project and spatial microsimulation software tool that estimates the prevalence of obesity at the local levels depending on demographic constraint variables (Edwards and Clarke, 2013). Recent evidence has emerged on the impact of car-dependent urban environments on inactive lifestyles and resulting poor health (these areas have been labelled 'obesogenic'). In this context, there is great potential for combining socio-demographic and environmental-geographic variables in a spatial microsimulation model. Using the same principles described by Tomintz, Clarke and Rigby (2008), the outputs of such a model could help target local interventions to tackle physical inactivity, maximising the benefits of public health initiatives.

3.3.2   Economic policy evaluation

The social-demographic distribution of impacts arising from economic policy evaluation is one of the most common applications of microsimulation (although the analysis in this area is often non-spatial). 'Social impact evaluation', where the impact of policy changes on different income and socio-demographic groups is explored, is a classic example of applied microsimulation research. Frequently these simulations are undertaken by government departments and focus on overall shifts in the population rather than spatial variability in the impacts. The EU–funded EUROMOD project, and software package of the same name, is the largest of these initiatives. The EUROMOD software is used by government analysts and research agencies in many countries to estimate the distributional impacts of policy reforms (Figure 3.2). The resulting research demonstrates that modelling work based on microsimulation can provide important new lines of evidence to inform national level policies (Avram et al., 2012).

Spatial microsimulation uses very similar techniques as those employed by EUROMOD and other economic microsimulation models, including probability-weighted random sampling of individual level data and aggregate level scenario development (Sutherland and Figari, 2013).

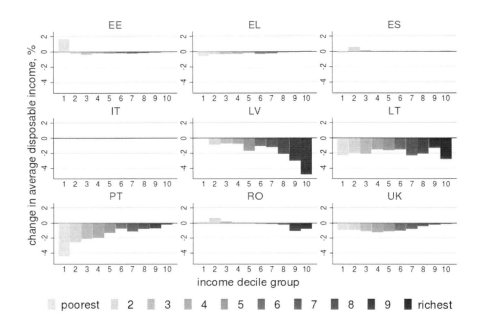

FIGURE 3.2
Output from the EUROMOD economic microsimulation model (Avram et al. 2014). Along the x axes is income group rising to the right. This means, for example, that Latvia (LV) has implemented progressive policies whereas Portugal (PT) has implemented regressive policies. Country acronyms from left to right stand for Estonia (EE), Greece (EL), Spain (ES), Italy (IT), Latvia (LV), Lithuania (LT), Portugal (PT), Romania (RO) and the UK.

The majority of microsimulation research for economic policy evaluation does not disaggregate the impacts over space, however. The estimation of variability at the *local level* is what differentiates spatial microsimulation models from economic microsimulation models, although the underlying methods are very similar. This book does not cover EUROMOD, focussing instead on spatially disaggregated microdata. However, there is great potential for future work to make EUROMOD more spatially enabled and to bring elements of EUROMOD into the methods outlined in the following chapters.

3.3.3 Transport

Transport modelling is a mature field that increasingly uses individual level data as the basis of analysis. Large scale models such as MATSim (`http://www.matsim.org/`) rely on spatial microdata to provide demand for travel and individual characteristics for origins and destinations. The same techniques are used in spatial microsimulation and transport modelling generating spatial microdata although in the transport literature, the process is referred to as *population synthesis* (Axhausen et al., 2011).

Generally, little attention is paid to this process of synthetic population generation in transport modelling because the focus is on movement of individuals rather than their characteristics. Distributional impacts are often overlooked in transport models (Lucas, 2012) and there is much potential to integrate spatial microsimulation with existing transport modelling methods.

An example of the potential uses of spatial microsimulation in transport models is illustrated in Figure 3.3. This shows the simulated commuting behaviour of 20 randomly selected individuals from a large scale spatial microdataset of Sheffield. Because the constraints used in this model included socio-demographic variables, each individual represented in the figure has a rich profile of characteristics associated with them. This analysis can provide new evidence about the likely winners and losers from very specific interventions such as a new bicycle path or bus route (Lovelace et al., 2014). As a result of the increased detail allowed by such methods there is much interest in spatial microsimulation for transport applications. Figure 3.3 also illustrates the potential for the output of spatial microsimulation to be used as an input into agent-based models (ABM).

A larger and more advanced illustration of the potential for spatial microsimulation in transport modelling work is described by Barthelemy and Toint (2015). This paper describes a stochastic model to allocate travel behaviours to a geo-located synthetic population of 10 million people, representing the entirety of the transport system in Belgium. By combining the synthetic microdata with an agent–based modelling approach (described in Chapter 12), Barthelemy and Toint (2015) are able to characterise a very large transport system in great detail.

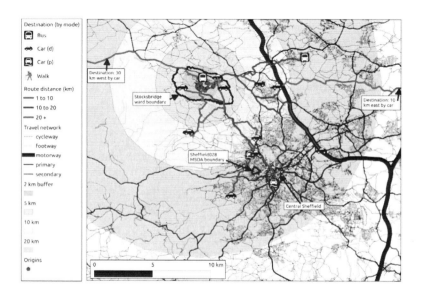

FIGURE 3.3
An illustration of spatial microdata in transport modelling. 20 people are
illustrated on the map as travelling to a range of destinations, specified based
on probability-weighted sampling from origin-destination tables (Lovelace et
al. 2014).

3.4 Assumptions

As with any simulation technique, spatial microsimulation is based on assumptions, some of which are unlikely to hold in all cases. This does not preclude spatial microsimulation in cases where the assumptions do not hold: "It is far better to foresee even without certainty than not to foresee at all", as Henri Poincaré put it (Barthélemy, 2014).

It is vital, however, that users of spatial microsimulation and 'consumers' of the resulting research understand that the results of spatial microsimulation are not *real* but a best estimate of the population in a given area. The danger is that if the assumptions are not understood, incorrect conclusions will result. It is therefore the duty of researchers using spatial microsimulation (and other techniques) to clearly state the assumptions on which the results depend on and the extent to which these assumptions can be expected to hold in practice. Roughly speaking there are four main assumptions underlying all spatial microsimulation models:

1. The individual level microdata are representative of the study area.
2. The target variable is dependent on the constraint variables and their interactions in a way that is relatively constant over space and time.
3. The relationships between the constraint variables are not spatially dependent.
4. The input microdataset and constraints are sufficiently rich and detailed to reproduce the full diversity of individuals and areas in the study region.

Obviously the real world is complex and many processes are spatially dependent, invalidating assumptions 2 and 3. The extent to which the relationships between variables can be deemed to be constant over space is often unknown. However, there are ways of checking the spatial dependency of relationships between multiply variables, not least Geographically Weighted Regression (GWR).

These limitations should be discussed at the outset of spatial microsimulation research, with reference to the input data. To see how spatial microsimulation simplifies the real world, the next chapter describes a hypothetical scenario where 33 inhabitants of an imaginary land are simulated and allocated to three zones based on a microdataset of only five individuals and two constraint variables.

3.5   Chapter summary

In this chapter we have defined what spatial microsimulation is and what it is not. Some research and real-world applications were described, with comments on areas for further work. The final section on assumptions underlying spatial microsimulation is in some ways the most important: It shows the importance of understanding the limitations associated with the method and the dangers of drawing conclusions from simulated data.

Part II

Generating spatial microdata

4

Data preparation

CONTENTS

Correctly loading, manipulating and assessing aggregate and individual level datasets is critical for effectively modelling real-world data. Getting the data into the right shape has the potential to make your models run quicker and increase the ease of modifying them, for example to include newly available input data. R is an accomplished tool for data reformatting, so it can be used as an integrated solution for both preparing the data and running the model.

In addition to providing a primer on data manipulation, the objects loaded in this chapter also provide the basis for Chapter 5, which covers the process of population synthesis in detail. Of course, the datasets you use in real applications will be much larger than those in SimpleWorld, but the concepts and commands will be largely the same.

Each project is different and data can be very diverse. For more on data cleaning and 'tidying', see Wickham (2014b). However, in most cases, the input datasets will be similar to the ones presented here: an individual level dataset consisting of categorical variables and aggregate level constraints containing categorical counts.

The process of loading, checking and preparing the input datasets for spatial microsimulation is generally a linear process, encapsulating the following stages, corresponding to the sections of this chapter:

- *Accessing the input data* (Section 4.1) provides advice on ways to collect and chose data.

- *Target and constraint variables* (Section 4.2) explains the different types of data you can have and how to define the targets and the constraints.
- *Loading input data* (Section 4.3) contains the R code to load the input data.
- *Subsetting to remove excess information* (Section 4.4) gives the R code to not consider all variables.
- *Re-categorising individual level variables* (Section 4.5) develops the R code to have pertinent categories in all variables.
- *Matching individual and aggregate level data names* (Section 4.6) makes the data names correspond to avoid problems when executing the spatial microsimulation.
- *'Flattening' the individual level data* (Section 4.7) explains a way to transform the data into a boolean matrices.

4.1   Accessing the input data

Before loading the spatial microdata, you must decide on the data needed for your research. Focus on "target variables" related to the research question will help decide on the constraint variables, as we will see below (Section 4.2). Selecting only on variables of interest will ensure you do not waste time thinking about and processing additional variables that will be of little use in the future. Regardless of the research question it is clear that the methodology depends on the availability of data. If the problem relates to variables that are simply unavailable at any level (e.g. hair length, if your research question related to the hairdressing industry), spatial microsimulation may be unsuitable. If, on the other hand, all the data is available at the geographical level of interest, spatial microsimulation may not be necessary.[1] Spatial microsimulation is useful when you have an intermediary amount of data available: geographically aggregated count and a non-spatial survey. In any case, you should have a clear idea about the range of available data on the topic of interest before embarking on spatial microsimulation.

As with most spatial microsimulation models, the input data in SimpleWorld consists of microdata — a non-geographical individual level dataset — and a constraint table which represents aggregate counts for a series of geographical zones. In some cases, you may have geographical information in your microdata, but not at the required level of detail. For example, you may have a variable on

---

[1] To explore geographical variability at a low spatial resolution, for example, the necessary data may be already available, as surveys often state which region each individual inhabits. Spatial microsimulation would only be necessary in this case if higher spatial resolution were needed.

the province of each individual but need its municipality, a lower geographical level. The input data can be accessed from the RStudio book project, as described in Chapter 2.

To ease reproducibility of the analysis when working with real data, it is recommended that the process begins with a copy of the *raw* dataset on your hard disc. Rather than modifying this file, modified ('cleaned') versions should be saved as separate files. This ensures that after any mistakes, one can always recover information that otherwise could have been lost and makes the project fully reproducible. In this chapter, a relatively clean and very tiny dataset from SimpleWorld is used, but we still create a backup of the original dataset. We will see in chapter 7 how to deal with larger and messier data, where being able to refer to the original dataset is more important. Here the focus is on the principles.

'Stripping down' the datasets so that they only contain the bare essential information will enable focus on the information that really matters. The input datasets in the example used for this chapter are already bare, but in real world surveys there may be literally hundreds of variables clogging up the analysis and your mind. It is good practice to remove excess data at the outset. Provided you have a good workflow, keeping all original data files unmodified for future reference, it will be easy to go back and add extra variables at a later stage. Following Occam's razor which favours simple solutions (Blumer et al. 1987), it is often best to start with a minimum of data and add complexity subsequently rather than vice versa.

Spatial microsimulation involves combining individual and aggregate level data. Each level should be given equal consideration when preparing the inputs for the modelling process. In many cases, the limiting factor for model fit will be the number of *linking variables*, variables shared between the individual and aggregate level datasets which are used to calculate the weights allocated to the individuals for each zone (see Glossary). These are also referred to as *constraint variables*, as they *constrain* the individual weights per zone (Ballas et al. 2005).

If there are no shared variables between the aggregate and individual level data, generating an accurate synthetic population is impossible. In this case, your only alternative is to consider only the marginal distribution of the variable and make a random selection with the distribution used as a probability. However, this implicitly assumes that there are no correlations between the available variables. If there are only a few shared variables (e.g. age, sex, employment status), your options are limited. Increasingly, however, there are many linking variables to choose from as questionnaires become broader and more available. In this case the choice of constraints becomes important: which and how many to use, their order and their relationship to the target variable should all be considered at the outset when choosing the input data and constraints.

4.2 Target and constraint variables

The geographically aggregated data should be the first consideration when deciding on the input data. This is because the geographical data is essential for making the model spatial: without good geographical data, there is no way to allocate the individuals to different geographical zones.

The first consideration for the constraint data is coverage: ideally close to 100% of each zone's total population should be included in the counts. Also, the survey must have been completed by a number of residents proportional to the total number of inhabitants in every geographical zone under consideration. It is important to recognise that many geographically aggregated datasets may be unsuitable for spatial microsimulation due to sample size: geographical information based on a small sample of a zone's population is not a good basis for spatial microsimulation.

To take one example, estimates of the *proportion* of people who do not get sufficient sleep in the Australian state of Victoria, from the 2011 VicHealth Indicators Survey (`http://data.aurin.org.au/dataset?q=sleep`), is not a suitable constraint variable.

Constraint variables should be integer counts, and should contain a sufficient number categories for each variable. Ideally, each relevant category and cross-tabulation should be represented. It is unusual to have a baby with a high degree qualification, but it may be useful to know there is an absence of such individuals. The sleep dataset meets neither of these criteria. It is based on a small sample of individuals (on average 300 per geographic zone, less than 1% of the total population). Also it is a binary variable, dividing people into 'adequate' and 'inadequate' sleep categories. Instead, geographical datasets from the 2011 Census should be used. These may contain fewer variables, but they will provide counts for different categories and have almost 100% coverage of the population.

To continue with the Australian example, imagine we are interested only in Census data at the SA2 level, the second smallest unit for the release of Census data. A browse of the available datasets on an online portal (`http://data.aurin.org.au/dataset?q=2011-08+SA2`) reveals that information is available on a wide range of topics, including industry of employment, education level and total personal weekly income (divided into 10 bands from $1 - $199 to $2000+). Which of these dozens of variables do we select of the analysis? It depends on the research question and the target variable or variables of interest.

If we are interested in geographic variability in economic inequality, for example, the income data would be first on the list to select. Additional variables would likely include level of education, type of employment and age, to discover the wider context of the problem. If the research question related to energy and sustainability, to take another example, variables related to distance travelled to work and housing would be of greater relevance. In each of these examples the *target variable* determines the selection of constraints. The target variable is the thing that we would like spatial microdata on, as illustrated by the following two research questions.

How does income inequality vary from place to place? To answer this question it is not sufficient to have aggregate level statistics (e.g. average household income). So income per person must be simulated for each zone, using spatial microsimulation. Sampling from a national population provides a more realistic distribution than simply assuming every person of a given income band earns a certain amount; this is clearly not the case as income distributions are not flat (they tend to have positive skew).

How does energy use vary over geographical space? This question is more complicated as there is no single variable on 'energy use' that is collected by statistical agencies at the aggregate, let alone individual, level. Rather, energy use is a *composite target variable* that is composed of energy use in transport, household heating and cooling and other things. For each type of energy use, the question must eventually be simplified to coincide with survey questions that are actually asked in Census surveys (e.g. 'where do you work?', from which distance travelled to work may be ascertained). Such considerations should guide the selection of aggregate level (generally Census-based) geographic count data. Of course, available datasets vary greatly from country to country so selection of appropriate datasets is highly context dependent.

The following considerations should inform the selection of individual level microdata:

- Linking variables: are there enough variables in the individual level data that are also present in the geographically aggregated constraints? Even when many linking variables are available, it is important to determine the fit between the categories in each. If age is reported in five-year bands in the aggregate level data, but only in 20-year bands in the individual level data, for example, this could be problematic for highly age-dependent research.
- Representiveness: is the sample population representative of the areas under investigation?
- Sample size and diversity: the input dataset must be sufficiently large and diverse to mitigate the *empty cell* problem.
- Target variables: are these, or proxies for them, included in the dataset?

In addition to these essential characteristics of the individual level dataset, there are qualities that are desirable but not required. These include:

- Continuity of the survey: will the survey be repeated on a regular basis into the future? If so, the analysis can form the basis of ongoing work to track change over time, perhaps as part of a *dynamic microsimulation model.*
- Geographic variables: although it is the purpose of spatial microsimulation to allocate geographic variables to individual level data, it is still useful to have some geographic information. In many national UK surveys, for example, the region (the coarsest geographical level) of the respondent is reported. This information can be useful in identifying the extent to which the difference between the geographic extent of the survey and microsimulation study area affects the results. This is an understudied area of knowledge where more work is needed.

In terms of the *number* of constraints that is appropriate there is no 'magic number' that is correct in every case. It is also important to note that the number of variables is not a particularly good measure of how well constrained the data will be. The total number of *categories* used to constrain the weights is in fact more important.

For example, a model constrained by two variables, each containing 10 categories (20 constraint categories in total), will be better constrained than a model constrained by 5 binary variables such as male/female, young/old etc. That is not to say that the former set of constraints is *better* (as emphasised above, that depends on the research question), simply that the weights will be more finely constrained.

A special type of constraint variable is **cross-tabulated** categories. This involves subdividing the categories in one variable by the categories of another. Cross-tabulated constraint variables provide a more detailed picture on not only the prevalence of particular categories, but the relationships between them. For the purposes of spatial microsimulation, cross-tabulated variables can be seen as a single variable.

If there are 5 age categories and 2 sex categories, this can be seen as a single constraint variable (age/sex) with 10 categories. Clearly in terms of retaining detailed local information (e.g. all young males moving out of a zone), cross-tabulated variables are preferable. This raises the question: what happens when a single variable (e.g age) is used in multiple cross-tabulated constraints (e.g. age/sex and age/income). In this case spatial microsimulation can still be used, but the result may not converge to a single answer because the gender distribution may be disrupted by the age/income constraint. Further work is needed to test this but from theory we can be sure: a 3-way cross-tabulated constraint (age/sex/income) would be ideal in this case because it provides more information than 2-way marginal distributions.

The level of detail within the linking variables is an important determinant of the fidelity of the resulting synthetic population. This can be measured in term of the number of linking variables (there are 2 in SimpleWorld: age and sex) and the more measure of the number of categories within all linking variables (there are 4 in SimpleWorld: male, female, young and old). This latter measure is preferable as it provides a closer indication of the 'bin size' used to categorise individuals during population synthesis. Still, the nature of both linking variables and their categories should be considered: an age variable containing 5 categories may look good on paper. However, if those categories are defined by the breaks `c(0, 5, 10, 15, 20, 90)`, the linking variable will not be effective at accurately recreating age distributions. More evenly distributed categories (such as `c(0, 20, 40, 60, 80, 110)` to continue the age example) are preferable.

(As an important aside for R learners, we use R syntax here to define the *breaks* instead of the more common '0 to 5', 6 to 10' age category notation because this is how numerical variables are best categorised in R. To see how this works, try entering the following into R's command line: `cut(x = c(20, 21, 90, 35), breaks = c(0, 20, 40, 60, 80, 110))`. The result is a `factor` variable as follows: `(0,20] (20,40] (80,110] (20,40]`. This output uses standard notation for defining bins of continuous data: the `(0, 20]` term means, in English, "any value from *greater than* zero, up to *and including* 20". In classic mathematical notation this reads $0 < x \leq 20$. Note the] symbol means 'including.'[2])

Even when measuring constraint variables by categories rather than the cruder measure of number variables, there is still no consensus about the most desirable number. Norman (1999) advocates including the "maximum amount of information in the constraints", implying that the more constraint categories the merrier. Harland et al. (2012) warns that over-constraining the model can cause problems. In any case, there is a clear link between the quality and quantity of constraint variables used in the model and the fidelity of the output microdata.

If constraint variables come from different sources, check the coherence of these datasets. In some cases the total number of individuals will not be consistent between constraint variables and the procedure will not work properly. This can happen if constraint variables are measured using different *base populations* or at different levels. Number of cars per household, for example, is usually collected at the household level so will contain lower counts than variables on individuals (e.g. age). A solution is to set all population totals equal to the most reliable constraint variable by multiplying values in each category by a fixed amount. However, caution should be taken when using this approach because it assumes that the relationship between categories is the same across all level or base populations. In the case of car ownership, where larger households are likely to own more cars, this assumption clearly would not hold; inferring

---

[2]This `[a,b)` category notation follows the International Organization for Standardization (ISO) 80000-2:2009 standard for mathematical notation: Square brackets indicate that the endpoint is included in the set, curved brackets indicate that the endpoint is not included.

individual level marginals from household level data would in this case lead to an underestimate of car availability.

After suitable constraint variables have been chosen — and remember that the constraints can be changed at a later stage to improve the model or for testing — the next stage is to load the data. Following the SimpleWorld example, we load the individual level dataset first. This is because individual level data from surveys is often more difficult to format. Individual level datasets are often larger and more complex than the constraint data, which are simply integer counts of different categories. Of course, it is possible that the data you have are not suitable for spatial microsimulation because they lack linking variables. We assume that you have already checked this. The checking process for the datasets used in this chapter is simple: both aggregate and individual level tables contain age (in continuous form in the microdata, as two categories in the aggregate data) and sex, so they can by combined. Loading the data involves transferring information from a local hard disc into R's *environment*, where it is available in memory.

4.3   Loading input data

Real-world individual level data are provided in many formats. These ultimately need to be loaded into R as a `data.frame` object. Note that there are many ways to load data into R, including `read.csv()` and `read.table()` from base R. Useful commands for loading proprietary data formats include `read_excel` and `read.spss()` from **readxl** and **foreign** packages, respectively.

More time consuming is cleaning the data and there are also many ways to do this. In the following example we present steps needed to load the data underlying the SimpleWorld example into R.[3] Some parts of the following explanation are specific to the example data and the use of Iterative Proportional Fitting (IPF) as a reweighting procedure (described in the next chapter). Combinatorial optimisation approaches, for example (as well as careful use of IPF methods, e.g. using the **mipfp** package), are resilient to differences in the number of individuals according to each constraint. However, the approach, functions and principles demonstrated will apply to the majority of real input datasets for spatial microsimulation. This section is quite specific to data preparation for spatial microsimulation; for a more general introduction to data formatting and cleaning methodology see Wickham (2014). To summarise

---

[3]All data and code to replicate the procedures outlined in the explicative chapters of the book are available publicly from the spatial-microsimulation-book GitHub repository, as described above. We recommend loading the input data and playing around with it.

this information, the last section of this chapter provides a check-list of items to ensure that the data has been adequately cleaned ready for the next phase.

In the SimpleWorld example, the individual level dataset is loaded from a 'plain text' (human readable) .csv file:

```
# Load the individual level data
ind <- read.csv("data/SimpleWorld/ind-full.csv")
class(ind) # verify the data type of the object
```

```
## [1] "data.frame"
```

```
ind # print the individual level data
```

```
##   id age sex income
## 1  1  59   m   2868
## 2  2  54   m   2474
## 3  3  35   m   2231
## 4  4  73   f   3152
## 5  5  49   f   2473
```

Constraint data are usually made available one variable at a time, so these are read in one file at a time:

```
con_age <- read.csv("data/SimpleWorld/age.csv")
con_sex <- read.csv("data/SimpleWorld/sex.csv")
```

We have loaded the aggregate constraints. As with the individual level data, it is worth inspecting each object to ensure that they make sense before continuing. Taking a look at `age_con`, we can see that this data set consists of 2 categories for 3 zones:

```
con_age
```

```
##   a0.49 a.50.
## 1     8     4
## 2     2     8
## 3     7     4
```

This tells us that there 12, 10 and 11 individuals in zones 1, 2 and 3, respectively, with different proportions of young and old people. Zone 2, for example, is heavily dominated by older people: there are 8 people over 50 whilst there are only 2 young people (under 49) in the zone.

Even at this stage there is a potential for errors to be introduced. A classic mistake with areal (geographically aggregated) data is that the order in which zones are loaded can change from one table to the next. The constraint data should therefore come with some kind of *zone id*, an identifying code. This usually consists of a unique character string or integer that allows the order of different datasets to be verified and for data linkage using attribute joins. Moreover, keeping the code associated with each administrative zone will subsequently allow attribute data to be combined with polygon shapes and visualised using GIS software.

If we're sure that the row numbers match between the age and sex tables (we are sure in this case), the next important test is to check the total populations of the constraint variables. Ideally both the *total* study area populations and *row totals* should match. If the *row totals* match, this is a very good sign that not only confirms that the zones are listed in the same order, but also that each variable is sampling from the same *population base*. These tests are conducted in the following lines of code:

```
sum(con_age)
```

```
## [1] 33
```

```
sum(con_sex)
```

```
## [1] 33
```

```
rowSums(con_age)
```

```
## [1] 12 10 11
```

```
rowSums(con_sex)
```

```
## [1] 12 10 11
```

```
rowSums(con_age) == rowSums(con_sex)
```

```
## [1] TRUE TRUE TRUE
```

The results of the previous operations are encouraging. The total population is the same for each constraint overall and for each area (row) for both constraints. If the total populations between constraint variables do not match (e.g. because the *population bases*[4] are different) this is problematic. Appropriate steps to normalise the errant constraint variables are described in the CakeMap chapter (7). This involves scaling the category totals so the totals are equal across all categories.

4.4   Subsetting to remove excess information

In the above code, `data.frame` objects containing precisely the information required for the next stage were loaded. More often, superfluous information will need to be removed from the data and subsets taken. It is worth removing superfluous variables early, to avoid over-complicating and slowing-down the analysis. For example, if `ind` had 100 variables of which only the 1st, 3rd and 19th were of interest, the following command could be used to update the object. Note that only the relevant variables, corresponding to the first, third and nineteenth columns, are retained:[5]

```
ind <- ind[, c(1, 3, 19)]
```

In the SimpleWorld dataset, only the `age` and `sex` variables are useful for reweighting: we can remove the others for the purposes of allocating individuals to zone. Note that it is important to keep track of individual Id's, to ensure individuals do not get mixed-up by a function that changes their order (`merge()`, for example, is a function that can cause havoc by changing the order of rows). Before removing the superfluous `income` variable, we will create a backup of `ind` that can be referred back to if necessary[6]:

```
ind_orig <- ind # store the original ind dataset
ind <- ind[, -4] # remove income variable
```

Although `ind` in this case is small, it will behave in the same way as larger datasets. Starting small is sensible, providing opportunities for testing subsetting syntax in R. It is common, for example, to take a subset of the working *population base*: those aged between 16 and 74 in full-time employment. Methods for doing this are provided in the Appendix (Section 13.3).

---

[4]see the Glossary for a description of the population base.

[5]An alternative way to remove excess variables is to use `NULL` assignment to remove columns. `ind$age <- NULL`, for example, would remove the age variable. The minus sign can also be used to remove specific rows or columns, as illustrated with the syntax `[, -4]` below.

[6]For example, when we will have a spatial microdataset replicating the individuals, we will be able to re-assign the income to each generated individual.

4.5 Re-categorising individual level variables

Before transforming the individual level dataset ind into a form that can be
compared with the aggregate level constraints, we must ensure that each dataset
contains the same information. It can be more challenging to re-categorise
individual level variables than to re-name or combine aggregate level variables,
so the former should usually be set first. An obvious difference between the
individual and aggregate versions of the age variable is that the former is of
type integer whereas the latter is composed of discrete bins: 0 to 49 and 50+.
We can categorise the variable into these bins using cut():

```
# Test binning the age variable
brks <- c(0, 49, 120) # set break points from 0 to 120 years
cut(ind$age, breaks = brks) # bin the age variable
```

```
## [1] (49,120] (49,120] (0,49]   (49,120] (0,49]
## Levels: (0,49] (49,120]
```

Note that the output of the above cut() command is correct, with individuals
binned into one of two bins, but that the labels are rather strange. To change
these category labels to something more readable for people who do not read
ISO standards for mathematical notation (most people!), we add another
argument, labels to the cut() function:

```
# Convert age into a categorical variable
labs <- c("a0_49", "a50+") # create the labels
cut(ind$age, breaks = brks, labels = labs)
```

```
## [1] a50+  a50+  a0_49 a50+  a0_49
## Levels: a0_49 a50+
```

The factor generated now has satisfactory labels: they match the column
headings of the age constraint, so we will save the result. (Note, we are not
losing any information at this stage because we have saved the original ind
object as ind_orig for future reference.)

```
# Overwrite the age variable with categorical age bands
ind$age <- cut(ind$age, breaks = brks, labels = labs)
```

Users should beware that cut results in a vector of class *factor*. This can cause
problems in subsequent steps if the order of constraint column headings is
different from the order of the factor labels, as we will see in the next section.

4.6 Matching individual and aggregate level data names

Before combining the newly re-categorised individual level data with the aggregate constraints, it is useful for the category labels to match up. This may seem trivial, but will save time in the long run. Here is the problem:

```
levels(ind$age)
```

```
## [1] "a0_49" "a50+"
```

```
names(con_age)
```

```
## [1] "a0.49" "a.50."
```

Note that the names are subtly different. To solve this issue, we can simply change the names of the constraint variable, after verifying they are in the correct order:

```
names(con_age) <- levels(ind$age) # rename aggregate variables
```

With both the age and sex constraint variable names now matching the category labels of the individual level data, we can proceed to create a single constraint object we label `cons`. We do this with `cbind()`:

```
cons <- cbind(con_age, con_sex)
cons[1:2, ] # display the constraints for the first two zones
```

```
##    a0_49 a50+ m f
## 1     8    4 6 6
## 2     2    8 4 6
```

4.7 'Flattening' the individual level data

We have made steps towards combining the individual and aggregate datasets and now only need to deal with 2 objects (`ind` and `cons`) which now share category and variable names. However, these datasets cannot possibly be compared because they measure very different things. The `ind` dataset records the value that each individual takes for a range of variables, whereas `cons` counts the number of individuals in different groups at the geographical level. These datasets are have different dimensions:

```
dim(ind)
```

```
## [1] 5 3
```

```
dim(cons)
```

```
## [1] 3 4
```

The above code confirms this: we have one individual level dataset comprising 5 individuals with 3 variables (2 of which are constraint variables and the other an ID) and one aggregate level constraint table called `cons`, representing 3 zones with count data for 4 categories across 2 variables.

The dimensions of at least one of these objects must change before they can be correctly easily compared. To do this we 'flatten' the individual level dataset. This means increasing its width so each column becomes a category name. This allows the individual data to be matched to the geographical constraint data. In this new dataset (which we label `ind_cat`, short for 'categorical'), each variable becomes a column containing Boolean numbers (either 1 or 0, representing whether the individual belongs to each category or not). Note that each row in `ind_cat` must contain a one for each constraint variable; the sum of every row in `ind_cat` should be equal to the number of constraints (this can be verified with `rowSums(ind_cat)`).

To undertake this 'flattening' process the `model.matrix()` function is used to expand each variable in turn. The result for each variable is a new matrix with the same number of columns as there are categories in the variable. Note that the order of columns is usually alphabetical: this can cause problems if the columns in the constraint tables are not ordered in this way. Knoblauch and Maloney (2012) provide a lengthier description of this flattening process.

The second stage is to use the `colSums()` function to take the sum of each column.[7]

---

[7] As we shall see in Section 5.2.3, only the former of these is needed if we use the **ipfp** package for re-weighting the data, but both are presented to enable a better understanding of how IPF works.

```
cat_age <- model.matrix(~ ind$age - 1)
cat_sex <- model.matrix(~ ind$sex - 1)[, c(2, 1)]

# Combine age and sex category columns into single data frame
(ind_cat <- cbind(cat_age, cat_sex)) # brackets -> print result
```

```
##    ind$agea0_49 ind$agea50+ ind$sexm ind$sexf
## 1             0           1        1        0
## 2             0           1        1        0
## 3             1           0        1        0
## 4             0           1        0        1
## 5             1           0        0        1
```

Note that second call to `model.matrix` is suffixed with `[, c(2, 1)]`. This is to swap the order of the columns: the column variables are produced from `model.matrix` is alphabetic, whereas the order in which the variables have been saved in the constraints object `cons` is `male` then `female`. Such subtleties can be hard to notice yet completely change one's results so be warned: the output from `model.matrix` will not always be compatible with the constraint variables.

To check that the code worked properly, let's count the number of individuals represented in the new `ind_cat` variable, using `colSums`:

```
colSums(ind_cat) # view the aggregated version of ind
```

```
## ind$agea0_49  ind$agea50+     ind$sexm     ind$sexf
##            2            3            3            2
```

```
ind_agg <- colSums(ind_cat) # save the result
```

The sum of both age and sex variables is 5 (the total number of individuals): it worked! Looking at `ind_agg`, it is also clear that the object has the same 'width', or number of columns, `cons`. This means that the individual level data can now be compared with the aggregate level data. We can check this by inspecting each object (e.g. via `View(ind_agg)`). A more rigorous test is to see if `cons` can be combined with `ind_agg`, using `rbind`:

```
rbind(cons[1,], ind_agg) # test compatibility of ind_agg and cons
```

```
##    a0_49 a50+ m f
## 1      8    4 6 6
## 2      2    3 3 2
```

If no error message is displayed on your computer, the answer is yes. This shows us a direct comparison between the number of people in each category of the constraint variables in zone and in the individual level dataset overall. Clearly, this is a very small example with only 5 individuals in total existing in `ind_agg` (the total for each constraint) and 12 in zone 1. We can measure the size of this difference using measures of *goodness of fit*. A simple measure is total absolute error (TAE), calculated in this case as `sum(abs(cons[1,]` `- ind_agg))`: the sum of the positive differences between cell values in the individual and aggregate level data.

The purpose of the *reweighting* procedure in spatial microsimulation is to minimise this difference (as measured in TAE above) by adding high weights to the most representative individuals.

4.8 Chapter summary

To summarise, we have learned about and implemented methods for loading and preparing input data for spatial microsimulation in this chapter. The following checklist outlines the main features of the input datasets to ensure they are ready for reweighting, covered in the next chapter:

1. Both constraint and target variables are loaded in R: in the former rows correspond to individuals; in the latter rows correspond to a spatial zone.
2. The categories of the constraint variables in the individual level dataset are identical to the column names of the constraint variables. (An example of the process needed to arrive at this state is the conversion of a continuous ages variable in the individual level dataset into a categorical variable to match the constraint data.)
3. The structure of the individual and aggregate level datasets must eventually be the same: the column names of `cons` and `ind_cat` in the above example are the categories of the constraint variables. `ind_cat` is the binary (or Boolean, containing 0s and 1s) version of the individual level dataset. This allows creation of an aggregated version of the individual level data that has the same dimensions (and is comparable with) the constraint data for each zone.
4. The total population of each zone is represented as the sum counts for each constraint variable.

Note that in much research requiring the analysis of complex data, the collection, cleaning and loading of the data can consume the majority of the project's time. This applies as much to spatial microsimulation as to any other data analysis task and is an essential stage before moving on the more exciting analysis and modelling. In the next chapter we progress to allocate weights to each individual in our sample, continuing with the example of SimpleWorld.

5

Population synthesis

CONTENTS

In this chapter we progress from loading and preparing the input data to running a spatial microsimulation model. The focus is on Iterative Proportional Fitting procedure (IPF), a simple, fast and widely-used method to allocate individuals to zones. Other methods can achieve the same result, as discussed in the first section of this chapter and as demonstrated in Chapter 6.

This chapter moves from theory to practice. It describes the process underlying the method of population synthesis (Section 5.2.1) before implementing the IPF algorithm in base R (Section 5.2.2). In the following sections two packages that automate IPF, **ipfp** and **mipfp**, are introduced (Section 5.2.3 and Section 5.2.4 respectively). The chapter also introduces the important concepts of *integerisation* and *expansion*. These methods are introduced towards the end of each package-specific section. The final section (5.6) compares the relative merits of **ipfp** and **mipfp**.

The SimpleWorld data, loaded in the previous chapter, is used as the basis of this chapter. Being small, simple and hopefully by now familiar, SimpleWorld facilitates understanding of the process on a 'human scale'. As always, we encourage experimentation with small datasets to avoid the worry of overloading your computer, before applying the same methods to larger and more complex data. Learning the basic principles and methods of spatial microsimulation in R is the focus of this chapter. Time spent mastering these basics will make subsequent steps much easier.

How representative each individual is of each zone is represented by their *weight* for that zone. Each weight links an individual to a zone. The number of weights is therefore equal to the number of zones multiplied by the number of individuals in the microdata. In terms of the SimpleWorld data loaded in the previous chapter we have, in R syntax, `nrow(cons)` zones and `nrow(ind)` individuals (typing those commands with the data loaded should confirm that there are 3 zones and 5 individuals in the input data for the SimpleWorld example). This means that `nrow(cons) * nrow(ind)` weights will be estimated (that is $3 * 5 = 15$ in SimpleWorld). The weights must begin with an initial value. We will create a matrix of ones. Thus, every individual is seen as equally representative of every zone:[1]

```
# Create the weight matrix.
# Note: relies on data from previous chapter.
weights <- matrix(data = 1, nrow = nrow(ind), ncol = nrow(cons))
dim(weights) # dimension of weight matrix: 5 rows by 3 columns
```

```
## [1] 5 3
```

The weight matrix links individual level data to aggregate level data. A weight matrix value of 0 in cell `[i,j]`, for example, suggests that the individual `i` is not representative of zone `j`. During the IPF procedure these weights are iteratively updated until they *converge* towards a single result: the final weights which create a representative population for each zone.[2]

5.1 Weighting algorithms

A wide range of methods can be used to allocate individuals to zones in spatial microsimulation. As with the majority of procedures for statistical analysis,

---

[1] Any initial weight values could be used here: initial weights do not affect the final weights after many iterations. The value of one is used as a sensible convention.

[2] Fienberg (1970) provides a geometrical proof that IPF converges to a single result, in the absence of *empty cells*.

there are *deterministic* and *stochastic* methods. The results of *deterministic* methods, such as IPF, never vary: no random or probabilistic numbers are used so the resulting weights will be the same every time. *Stochastic* methods such as *simulated annealing*, on the other hand, use some random numbers.

In the literature, the divide between stochastic and deterministic approaches is usually mapped onto a wider distinction between *reweighting* and *combinatorial optimisation* methods. Reweighting methods generally calculate non-integer weights for every individual-zone combination. Combinatorial optimisation methods generally work by randomly allocating individuals from the microdata survey into each zone, one-by-one, and re-calculating the goodness-of-fit after each change. If the fit between observed and simulated results improves after an individual has been 'imported' into the zone in question, the individual will stay. If the fit deteriorates, the individual will be removed from the zone and the impact of switching a different individual into the zone is tested.

This distinction between reweighting of fractional weights and combinatorial optimisation algorithms is important: combinatorial optimisation methods result in whole individuals being allocated to each zone whereas reweighting strategies result in fractions of individuals being allocated to each zone. The latter method means that individual i could have a weight of 0.223 (or any other positive real number) for zone j. Of course, this sounds irrational: a quarter of a person cannot possibly exist: either she is in the zone or she is not!

However, the distinction between combinatorial optimisation and reweighting approaches to creating spatial microdata is not as clear cut as it may at first seem. As illustrated in Figure 5.1, fractional weights generated by reweighting algorithms such as IPF can be converted into integer weights (via *integerisation*).

Through the process of *expansion*, the integer weight matrix produced by integerisation can be converted into the final spatial microdata output stored in 'long' format represented in the right-hand box of Figure~5.1. Thus the combined processes of integerisation and expansion allow weight matrices to be translated into the same output format that combinatorial optimisation algorithms produce directly. In other words fractional weighting is interchangeable with combinatorial optimisation approaches for population synthesis.

The reverse process is also possible: synthetic spatial microdata generated by combinatorial optimisation algorithms can be converted back in to a more compact weight matrix in a step that we call *compression*. This integer weight has the same dimensions as the integer matrix generated through integerisation described above.

Integerisation, expansion and compression procedures allow fractional weighting and combinatorial optimisation approaches for population synthesis to be seen as essentially the same thing.

This equivalence between different methods of population synthesis is the reason we have labelled this section *weighting algorithms*: combinatorial optimisation approaches to population synthesis can be seen as a special case of fractional weighting and vice versa. Thus all deterministic and stochastic (or weighting and combinatorial optimisation) approaches to the generation of spatial microdata can be seen as different methods, algorithms, for allocating weights to individuals. Individuals representative of a zone will be given a high weight (which is equivalent to being replicated many times in combinatorial optimisation). Individuals who are rare in a zone will be given a low weight (or not appear at all, equivalent to a weight of zero). Later in this chapter we demonstrate functions to translate between the 'weight matrix' and 'long spatial microdata' formats generated by each approach.

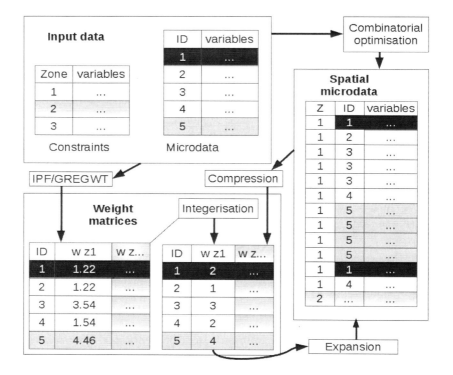

FIGURE 5.1

Schematic of different approaches for the creation of spatial microdata encapsulating stochastic combinatorial optimisation and deterministic reweighting algorithms such as IPF. Note that integerisation and 'compression' steps make the results of the two approaches interchangeable, hence our use of the term 'reweighting algorithm' to cover all methods for generating spatial microdata.

The concept of weights is critical to understanding how population synthesis generates spatial microdata. To illustrate the point imagine a parallel SimpleWorld, in which we have no information about the characteristics of its inhabitants, only the total population of each zone. In this case we could only assume that the distribution of characteristics found in the sample is representative of the distribution of the whole population. Under this scenario, individuals would be chosen at random from the sample and allocated to zones at random and the distribution of characteristics of individuals in each zone would be asymptotically the same as (tending towards) the microdata.

In R, this case can be achieved using the `sample()` command. We could use this method to randomly allocate the 5 individuals of the microdata to zone 1 (which has a population of 12) with the following code:

```
# set the seed for reproducibility
set.seed(1)
# create selection
sel <- sample(x = 5, size = 12, replace = T)
ind_z1 <- ind_orig[sel, ]
head(ind_z1, 3)
```

```
##      id age sex income
## 2     2  54   m   2474
## 2.1   2  54   m   2474
## 3     3  35   m   2231
```

Note the use of `set.seed()` in the above code to ensure the results are reproducible. It is important to emphasise that without 'setting the seed' (determining the starting point of the random number sequence) the results will change each time the code is run. This is because `sample()` (along with other stochastic functions such as `runif()` and `rnorm()`) is probabilistic and its output therefore depends on a random number generator (RNG).[3]

The *reweighting* methods consists in adding a weight to each individual for the zone. This method is good if we have a representative sample of the zone and the minority of the population is included in it. In contrary, if we have in the individual level only the majority of the population and for example, we have not an old man still working, this kind of individual will not appear in the

---

[3]Without setting the seed, the results will change each time. In addition, changing the value inside the brackets of `set.seed()` will result in a different combination of individuals being selected for each new number — test this out in your code. This happens because the method relies on *pseudo random numbers* to select values probabilistically and `set.seed()` specifies where the random number sequence should begin, ensuring reproducibility. We must really trust the random function used. Note that it is impossible for a computer to choose *entirely* random numbers, so algorithms to generate *pseudo-random numbers* have been developed. See the documentation provided by `?set.seed` for more information.

final data. A proposal to avoid this is to either modify the sample to include those outliers or use a different approach such as a *genetic algorithm* to allow mixing some individual level variables to create a new individuals (mutation). We are not aware of the latter technique being used at present, although it could be used in future microsimulation work. An other useful alternative to solve this issue is given by the *tabu-search algorithm*, an heuristic for integer programming method. (Glover 1986, 1989, 1990 and Glover and Laguna, 1997). This heuristic can be used to improve an initial solution produced by the IPF.

In the rest of this chapter, we develop an often used reweighting algorithm : the Iterative Proportional Fitting (IPF). When reweighting, three steps are always necessary. First, you generate a weight matrix containing fractional numbers, then you integerise it. When knowing the number of times each individual needs to be replicated, the expansion step calculates the final dataset, which contains a row per individual in the whole population and a zone for each person. The process is explained in theory and then demonstrated with example code in R.

5.2 Iterative Proportional Fitting

This section begins with the theory (and intuition) under the IPF algorithm. Then, the R code to run the method is explained. However, in R, it is often unnecessary to write the whole code. Indeed, packages including functions to perform an IPF exist and are optimised. The **ipfp** and **mipfp** packages are the subjects of the two last sub-sections of this section.

5.2.1 IPF in theory

The most widely used and mature *deterministic* method to allocate individuals to zones is iterative proportional fitting (IPF). IPF is mature, fast and has a long history: it was demonstrated by Deming and Stephan (1940) for estimating internal cells based on known marginals. IPF involves calculating a series of non-integer weights that represent how representative each individual is of each zone. This is *reweighting*.

Regardless of implementation method, IPF can be used to allocate individuals to zones through the calculation of *maximum likelihood* values for each zone-individual combination represented in the weight matrix. This is considered as the most probable configuration of individuals in these zones. IPF is a method of *entropy maximisation* (Cleave et al. 1995). The *entropy* is the number of configurations of the underlying spatial microdata that could result in the same marginal counts. For in-depth treatment of the mathematics and theory underlying IPF, interested readers are directed towards Fienberg (1979), Cleave et al. (1995) and an excellent recent review of methods for maximum likelihood estimation (Fienberg and Rinaldo 2007). For the purposes of this book, we provide a basic overview of the method for the purposes of spatial microsimulation.

In spatial microsimulation, IPF is used to allocate individuals to zones. The subsequent section implements IPF to create *spatial microdata* for SimpleWorld, using the data loaded in the previous chapter as a basis. An overview of the paper from the perspective of transport modelling is provided by Pritchard and Miller (2012).

Such as with the example of SimpleWorld, each application has a matrix `ind` containing the categorical value of each individual. `ind` is a two dimensional array (a matrix) in which each row represents an individual and each column a variable. The value of the cell `ind(i,j)` is therefore the category of the individual i for the variable j.

A second array containing the constraining count data `cons` can, for the purpose of explaining the theory, be expressed in three dimensions, which we will label `cons_t`: `cons_t(i,j,k)` is the number of individuals corresponding to the marginal for the zone 'i', in the variable 'j' for the category 'k'. For example, 'i' could be a municipality, 'j' the gender and 'k' the female. Element '(i,j,k)' is the total number of women in this municipality according to the constraint data.

The IPF algorithm will proceed zone per zone. For each zone, each individual will have a weight of 'representatively' of the zone. The weight matrix will then have the dimension 'number of individual x number of zone'. 'w(i,j,t)' corresponds to the weight of the individual 'i' in the zone 'j' (during the step 't'). For the zone 'z', we will adapt the weight matrix to each constraint 'c'. This matrix is initialized with a full matrix of 1 and then, for each step 't', the formula can be expressed as:

$$w(i,z,t+1) = w(i,z,t) * \frac{cons\_t(z,c,ind(i,c))}{\sum_{j=1}^{n_{ind}} w(j,z,t) * I(ind(j,c) = ind(i,c))}$$

where the 'I(x)' function is the indicator function which value is 1 if x is true and 0 otherwise. We can see that 'ind(i,c)' is the category of the individual 'i' for the variable 'c'. The denominator corresponds to the sum of the actual weights of all individuals having the same category in this variable as 'i'. We simply redistribute the weights so that the data follows the constraint concerning this variable.

5.2.2   IPF in R

In the subsequent examples, we use IPF to allocate individuals to zones in
SimpleWorld, using the data loaded in the previous chapter as a basis. IPF is
mature, fast and has a long history.

The IPF algorithm can be written in R from scratch, as illustrated in
Lovelace (2014), and as taught in the smsim-course on-line tutorial (`https://github.com/Robinlovelace/spatial-microsim-book`). We will refer to
this implementation of IPF as 'IPFinR'. The code described in this tutorial
'hard-codes' the IPF algorithm in R and must be adapted to each new application (unlike the generalized 'ipfp' approach, which works unmodified on
any reweighting problem). IPFinR works by saving the weight matrix after
every constraint for each iteration. We here develop first IPFinR to give you
an idea of the algorithm. Without reading the *source code* in the **ipfp** and
mipfp packages, available from CRAN, they can seem like "black boxes" which
magically generate answers without explanation of *how* answer was created.

By the end of this section you will have generated a weight matrix representing
how representative each individual is of each zone. The algorithm operates
zone per zone and the weight matrix will be filled in column per column. For
convenience, we begin by assigning some of the basic parameters of the input
data to intuitive names, for future reference.

```
# Create intuitive names for the totals
n_zone <- nrow(cons) # number of zones
n_ind <- nrow(ind) # number of individuals
n_age <-ncol(con_age) # number of categories of "age"
n_sex <-ncol(con_sex) # number of categories of "sex"
```

The next code chunk creates the R objects needed to run IPFinR. The code
creates the weight matrix (`weights`) and the marginal distribution of individuals in each zone (`ind_agg0`). Thus first we create an object (`ind_agg0`), in
which rows are zones and columns are the different categories of the variables.
Then, we duplicate the weight matrix to keep in memory each step.

```
# Create initial matrix of categorical counts from ind
weights <- matrix(data = 1, nrow = nrow(ind), ncol = nrow(cons))
# create additional weight objects
weights1 <- weights2 <- weights
ind_agg0 <- t(apply(cons, 1, function(x) 1 * ind_agg))
colnames(ind_agg0) <- names(cons)
```

IPFinR begins with a couple of nested 'for' loops, one to iterate through
each zone (hence `1:n_zone`, which means "from 1 to the number of zones in

the constraint data") and one to iterate through each category within the constraints (0–49 and 50+ for the first constraint), using the `cons` and `ind` objects loaded in the previous chapter.

```
# Assign values to the previously created weight matrix
# to adapt to age constraint
for(j in 1:n_zone){
  for(i in 1:n_age){
    index <- ind_cat[, i] == 1
    weights1[index, j] <- weights[index, j] * con_age[j, i]
    weights1[index, j] <- weights1[index, j] / ind_agg0[j, i]
    }
  print(weights1)
}
```

```
##              [,1] [,2] [,3]
## [1,] 1.333333    1    1
## [2,] 1.333333    1    1
## [3,] 4.000000    1    1
## [4,] 1.333333    1    1
## [5,] 4.000000    1    1
##              [,1]       [,2] [,3]
## [1,] 1.333333 2.666667    1
## [2,] 1.333333 2.666667    1
## [3,] 4.000000 1.000000    1
## [4,] 1.333333 2.666667    1
## [5,] 4.000000 1.000000    1
##              [,1]       [,2]       [,3]
## [1,] 1.333333 2.666667 1.333333
## [2,] 1.333333 2.666667 1.333333
## [3,] 4.000000 1.000000 3.500000
## [4,] 1.333333 2.666667 1.333333
## [5,] 4.000000 1.000000 3.500000
```

The above code updates the weight matrix by multiplying each weight by a coefficient to respect the desired proportions (age constraints). These coefficients are the divisions of each cell in the census constraint (`con_age`) by the equivalent cell aggregated version of the individual level data. In the first step, all weights were initialised to 1 and the new weights equal the coefficients. Note that the two lines changing `weights1` could be joined to form only one line (we split it into two lines to respect the margins of the book). The weight matrix is critical to the spatial microsimulation procedure because it describes how representative each individual is of each zone. To see the weights that have been allocated to individuals to populate, for example zone 2, you would query the second column of the weights: `weights1[, 2]`. Conversely, to see the weight allocated for individual 3 for each for the 3 zones, you need to look at the 3rd column of the weight matrix: `weights1[3, ]`.

Note that we asked R to write the resulting matrix after the completion of each zone. As said before, the algorithm proceeds zone per zone and each column of the matrix corresponds to a zone. This explains why the matrix is filled in per column. We can now verify that the weights correspond to the application of the theory seen before.

For the first zone, the age constraint was to have 8 people under 50 years old and 4 over this age. The first individual is a man of 59 years old, so over 50. To determine the weight of this person inside the zone 1, we multiply the actual weight, 1, by a ratio with a numerator corresponding to the number of persons in this category of age for the constraint, here 4, and a denominator equal to the sum of the weights of the individual having the same age category. Here, there are 3 individuals of more than 50 years old and all weights are 1 for the moment. The new weight is

$$1 * \frac{4}{1 + 1 + 1} = 1.33333$$

We can verify the other weights with a similar reasoning. Now that we explained the whole process under IPF, we can understand the origin of the name 'Iterative Proportional Fitting'.

Thanks to the weight matrix, we can see that individual 3 (whose attributes can be viewed by entering `ind[3, ]`) is young and has a comparatively low weight of 1 for zone two. Intuitively this makes sense because zone 2 has only 2 young adult inhabitants (see the result of `cons[2,]`) but 8 older inhabitants. The reweighting stage is making sense. Note also that the weights generated are fractional; see Section 5.3 for methods of converting fractional weights into integer weights to generate a synthetic small-area population ready for agent-based modelling applications.

The next step in IPF, however, is to re-aggregate the results from individual level data after they have been reweighted. For the first zone, the weights of each individual are in the first column of the weight matrix. Moreover, the characteristics of each individual are inside the matrix `ind_cat`.

When multiplying `ind_cat` by the first column of `weights1` we obtain a vector, the values of which correspond to the number of people in each category for zone 1. To aggregate all individuals for the first zone, we just sum the values in each category. The following `for` loop re-aggregates the individual level data, with the new weights for each zone:

```
# Create additional ind_agg objects
ind_agg2 <- ind_agg1 <- ind_agg0 * NA
```

```
# Assign values to the aggregated data after con 1
for(i in 1:n_zone){
  ind_agg1[i, ] <- colSums(ind_cat * weights1[, i])
}
```

If you are new to IPF, congratulations: you have just reweighted your first individual level dataset to a geographical constraint (the age) and have aggregated the results. At this early stage it is important to do some preliminary checks to ensure that the code is working correctly. First, are the resulting populations for each zone correct? We check this for the first constraint variable (age) using the following code (to test your understanding, try to check the populations of the unweighted and weighted data for the second constraint — sex):

```
rowSums(ind_agg1[, 1:2]) # the simulated populations in each zone
```

```
## [1] 12 10 11
```

```
rowSums(cons[, 1:2]) # the observed populations in each zone
```

```
## [1] 12 10 11
```

The results of these tests show that the new populations are correct, verifying the technique. But what about the fit between the observed and simulated results after constraining by age? We will cover goodness of fit in more detail in subsequent sections. For now, suffice to know that the simplest way to test the fit is by using the `cor` function on a 1d representation of the aggregate level data:

```
vec <- function(x) as.numeric(as.matrix(x))
cor(vec(ind_agg0), vec(cons))
```

```
## [1] -0.3368608
```

```
cor(vec(ind_agg1), vec(cons))
```

```
## [1] 0.628434
```

The point here is to calculate the correlation between the aggregate actual data and the constraints. This value is between -1 and 1 and in our case, the best fit will be 1, meaning that there is a perfect correlation between our data and the constraints. Note that as well as creating the correct total population for each zone, the new weights also lead to much better fit. To see how this has worked, let's look at the weights generated for zone 1:

```
weights1[, 1]
```

```
## [1] 1.333333 1.333333 4.000000 1.333333 4.000000
```

The results mean that individuals 3 and 5 have been allocated a weight of 4 whilst the rest have been given a weight of 4/3. Note that the total of these weights is 12, the population of zone 1. Note also that individuals 3 and 5 are in the younger age group (verify this with `ind$age[c(3,5)]`) which are more commonly observed in zone 1 than the older age group:

```
cons[1, ]
```

```
##    a0_49 a50+ m f
## 1      8    4 6 6
```

Note there are 8 individuals under 50 years old in zone 1, but only 2 individuals with this age in the individual level survey dataset. This explains why the weight allocated to these individuals was 4: 8 divided by 2 = 4.

So far we have only constrained by age. This results in aggregate level results that fit the age constraint but not the sex constraint (Figure 5.2). The reason for this should be obvious: weights are selected such that the aggregated individual level data fits the age constraint perfectly, but no account is taken of the sex constraint. This is why IPF must constrain for multiple constraint variables.

To constrain by sex, we simply repeat the nested `for` loop demonstrated above for the sex constraint. This is implemented in the code block below.

```
for(j in 1:n_zone){
  for(i in 1:n_sex + n_age){
    index <- ind_cat[, i] == 1
    weights2[index, j] <- weights1[index, j] * cons[j , i] /
      ind_agg1[j, i]
    }
}
```

Again, the aggregate values need to be calculated in a `for` loop over every zone. After the first constraint fitting, the weights for zone 1 were:

$$(\frac{4}{3}, \frac{4}{3}, 4, \frac{4}{3}, 4)$$

We can explain theoretically explain the weights for zone 1 after the second fitting. For the first individual, its actual weight is $\frac{4}{3}$ and he is a male. In zone 1, the constraint is to have 6 men. The three first individuals are men, so the new weight for this person in this zone is

$$weights2[1,1] = \frac{4}{3} * \frac{6}{\frac{4}{3} + \frac{4}{3} + 4} = \frac{6}{5} = 1.2$$

With an analogous reasoning, we can find all weights in *weights2*:

```
weights2
```

```
##        [,1]      [,2]        [,3]
## [1,]   1.2 1.6842105 0.6486486
## [2,]   1.2 1.6842105 0.6486486
## [3,]   3.6 0.6315789 1.7027027
## [4,]   1.5 4.3636364 2.2068966
## [5,]   4.5 1.6363636 5.7931034
```

Note that the final value is calculated by multiplying by `ind_cat` *and* `weights2`.

```
for(i in 1:n_zone){
  ind_agg2[i, ] <- colSums(ind_cat * weights2[, i])
}
```

```
ind_agg2
```

```
##            a0_49      a50+ m f
## [1,] 8.100000 3.900000 6 6
## [2,] 2.267943 7.732057 4 6
## [3,] 7.495806 3.504194 3 8
```

Note that even after constraining by age and sex, there is still not a perfect fit between observed and simulated cell values (Figure 5.2). After constraining only by age, the simulated cell counts for the sex categories are far from the observed, whilst the cell counts for the age categories fit perfectly. On the other hand, after constraining by age *and* sex, the fit is still not perfect. This time the age categories do not fit perfectly, whilst the sex categories fit perfectly. Inspect `ind_agg1` and `ind_agg2` and try to explain why this is. The concept of IPF is to repeat this procedure several times. Thus, each iteration contains a re-constraining for each variable. In our case, the name iteration 1.1 and 1.2 describe the first iteration constrained only by age and then by sex, respectively. The *overall fit*, combining age and sex categories, has improved greatly from iteration 1.1 to 1.2 (from $r = 0.63$ to $r = 0.99$). In iteration 2.1 (in which we constrain by age again) the fit improves again.

So, each time we reweight to a specific constraint, the fit of this constraint is perfect, because, as seen in theory, it is a proportional reallocation. Then, we repeat for another constraint and the first one can diverge from this perfect fit. However, when operating this process several times, we always refit to the next constraint and we can converge to a unique weight matrix.

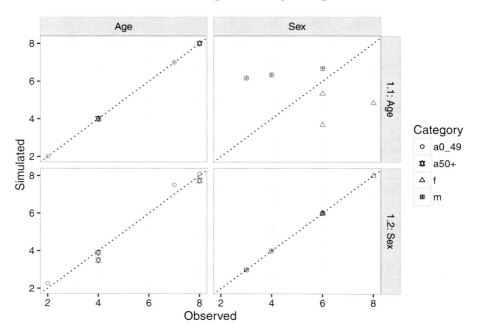

FIGURE 5.2

Fit between observed and simulated values for age and sex categories (column facets) after constraining a first time by age and sex constraints (iterations 1.1 and 1.2, plot rows). The dotted line in each plot represents perfect fit between the simulated and observed cell values. The overall fit in each case would be found by combining the left and right-hand plots. Each symbol corresponds to a category and each category has a couple (observed, simulated) for each zone.

These results show that IPF requires multiple *iterations* before converging on a single weight matrix. It is relatively simple to iterate the procedure illustrated in this section multiple times, as described in the smsim-course tutorial. However, for the purposes of this book, we will move on now to consider implementations of IPF that automate the fitting procedure and iteration process. The aim is to make spatial microsimulation as easy and accessible as possible.

The advantage of hard-coding the IPF process, as illustrated above, is that it helps understand how IPF works and aids the diagnosis of issues with the reweighting process as the weight matrix is re-saved after every constraint and iteration. However, there are more computationally efficient approaches to IPF. To save computer and researcher time, we use in the next sections R packages which implement IPF without the user needing to *hard code* each iteration in R: **ipfp** and **mipfp**. We will use each of these methods to generate fractional weight matrices allocating each individual to zones. After following this reweighting process with **ipfp**, we will progress to integerisation: the process of converting the fractional weights into integers. This process creates a final contingency table, which is used to generate the final population. This last step is called the expansion.

5.2.3   IPF with ipfp

IPF runs much faster and with less code using the **ipfp** package than in pure R. The `ipfp` function runs the IPF algorithm in the C language, taking aggregate constraints, individual level data and an initial weight vector (`x0`) as inputs:

```
library(ipfp) # load ipfp library after install.packages("ipfp")
cons <- apply(cons, 2, as.numeric) # to 1d numeric data type
ipfp(cons[1,], t(ind_cat), x0 = rep(1, n_ind)) # run IPF
```

```
## [1] 1.227998 1.227998 3.544004 1.544004 4.455996
```

It is impressive that the entire IPF process, which takes dozens of lines of code in base R (the IPFinR method), can been condensed into two lines: one to convert the input constraint dataset to `numeric`[4] and one to perform the IPF operation itself. The whole procedure is hiding behind the function that is created in C and optimised. So, it is like a magic box where you put your data and that returns the results. This is a good way to execute the algorithm, but one must pay attention to the format of the input argument to use the function correctly. To be sure, type on R `?ipfp`.

Note that although we did not specify how many iterations to run, the above command ran the default of `maxit = 1000` iterations, despite convergence happening after 10 iterations. Note that 'convergence' in this context means that the norm of the matrix containing the difference between two consecutive iterations reaches the value of `tol`, which can be set manually in the function (e.g. via `tol = 0.0000001`). The default value of `tol` is `.Machine$double.eps`, which is a very small number indeed: 0.000000000000000222, to be precise.

---

[4]The integer data type fails because C requires `numeric` data to be converted into its *floating point* data class.

The number of iterations can also be set by specifying the `maxit` argument (e.g. `maxit = 5`). The result after each iteration will be printed if we enable the verbose argument with `verbose = TRUE`. Note also that these arguments can be referred to lazily using only their first letter: `verbose` can be referred to as `v`, for example, as illustrated below (not all lines of R output are shown):

```
ipfp(cons[1,], t(ind_cat), rep(1, n_ind), maxit = 20, v = T)
```

```
## iteration 0:    0.141421
## iteration 1:    0.00367328
## iteration 2:    9.54727e-05
## ...
## iteration 9:    4.96507e-16
## iteration 10:   4.96507e-16
```

To be clear, the `maxit` argument in the above code specifies the maximum number of iterations and setting `verbose` to `TRUE` (the default is `FALSE`) instructed R to print the result. Being able to control R in this way is critical to mastering spatial microsimulation with R.

The numbers that are printed in the output from R correspond to the 'distance' between the previous and actual weight matrices. When the two matrices are equal, the algorithm has converged and the distance will approach 0. If the distance falls below the value of `tol`, the algorithm stops. Note that when calculating, the computer makes numerical approximations of real numbers. For example, when calculating the result of $\frac{4}{3}$, the computer cannot save the infinite number of decimals and truncates the number.

Due to this, we rarely reach a perfect 0 but we assume that a result that is very close to zero is sufficient. Usually, we use the precision of the computer that is of order 10^{-16} (on R, you can display the precision of the machine by typing `.Machine$double.eps`).

Notice also that for the function to work a *transposed* (via the `t()` function) version of the individual level data (`ind_cat`) was used. This differs from the `ind_agg` object used in the pure R version. To prevent having to transpose `ind_cat` every time `ipfp` is called, we save the transposed version:

```
ind_catt <- t(ind_cat) # save transposed version of ind_cat
```

Another object that can be saved prior to running `ipfp` on all zones (the rows of `cons`) is `rep(1, nrow(ind))`, simply a series of ones - one for each individual. We will call this object `x0` as its argument name representing the starting point of the weight estimates in `ipfp`:

```
x0 <- rep(1, n_ind) # save the initial vector
```

To extend this process to all three zones we can wrap the line beginning `ipfp(...)` inside a `for` loop, saving the results each time into a weight variable we created earlier:

```
weights_maxit_2 <- weights # create a copy of the weights object
for(i in 1:ncol(weights)){
  weights_maxit_2[,i] <- ipfp(cons[i,], ind_catt, x0, maxit = 2)
}
```

The above code uses `i` to iterate through the constraints, one row (zone) at a time, saving the output vector of weights for the individuals into columns of the weight matrix. To make this process even more concise (albeit less clear to R beginners), we can use R's internal `for` loop, `apply`:

```
weights <- apply(cons, MARGIN = 1, FUN =
    function(x) ipfp(x, ind_catt, x0, maxit = 20))
```

In the above code R iterates through each row (hence the second argument `MARGIN` being 1, `MARGIN = 2` would signify column-wise iteration). Thus `ipfp` is applied to each zone in turn, as with the `for` loop implementation. The speed savings of writing the function with different configurations are benchmarked in 'parallel-ipfp.R' in the 'R' folder of the book project directory. This shows that reducing the maximum iterations of `ipfp` from the default 1000 to 20 has the greatest performance benefit.[5] To make the code run faster on large datasets, a parallel version of `apply` called `parApply` can be used. This is also tested in 'parallel-ipfp.R'.

Take care when using `maxit` as the only stopping criteria as it is impossible to predict the number of iterations necessary to achieve the desired level of fit for every application. So, if your argument is too big you will needlessly lose time, but if it is too small, the result will not converge.

Using `tol` alone is also hazardous. Indeed, if you have iteratively the same matrices, but the approximations on the distance is a number bigger than your argument, the algorithm will continue. That's why we use `tol` and `maxit` in tandem. Default values for `maxit` and `tol` are 1000 a very small number defined in R as `.Machine$double.eps` respectively.

It is important to check that the weights obtained from IPF make sense. To do this, we multiply the weights of each individual by rows of the `ind_cat`

---

[5]These tests also show that any speed gains from using `apply` instead of `for` are negligible, so whether to use `for` or `apply` can be decided by personal preference.

matrix, for each zone. Again, this can be done using a for loop, but the apply method is more concise:

```
ind_agg <- t(apply(weights, 2, function(x) colSums(x * ind_cat)))
colnames(ind_agg) <- colnames(cons) # make the column names equal
```

As a preliminary test of fit, it makes sense to check a sample of the aggregated weighted data (`ind_agg`) against the same sample of the constraints. Let's look at the results (one would use a subset of the results, e.g. `ind_agg[1:3, 1:5]` for the first five values of the first 3 zones for larger constraint tables found in the real world):

```
ind_agg
```

```
##        a0_49 a50+ m f
## [1,]       8    4 6 6
## [2,]       2    8 4 6
## [3,]       7    4 3 8
```

```
cons
```

```
##        a0_49 a50+ m f
## [1,]       8    4 6 6
## [2,]       2    8 4 6
## [3,]       7    4 3 8
```

This is a good result: the constraints perfectly match the results generated using IPF, at least for the sample. To check that this is due to the `ipfp` algorithm improving the weights with each iteration, let us analyse the aggregate results generated from the alternative set of weights, generated with only 2 iterations of IPF:

```
# Update ind_agg values, keeping col names (note '[]')
ind_agg[] <- t(apply(weights_maxit_2, MARGIN = 2,
  FUN = function(x) colSums(x * ind_cat)))
ind_agg[1:2, 1:4]
```

```
##             a0_49      a50+ m f
## [1,] 8.002597 3.997403 6 6
## [2,] 2.004397 7.995603 4 6
```

Clearly the final weights after 2 iterations of IPF represent the constraint variables well, but do not match perfectly except in the second constraint. This shows the importance of considering number of iterations in the reweighting stage — too many iterations can be wasteful, too few may result in poor results. To reiterate, 20 iterations of IPF are sufficient in most cases for the results to converge towards their final level of fit. More sophisticated ways of evaluating model fit are presented in Section 8. As mentioned at the beginning of this chapter there are alternative methods for allocating individuals to zones. These are discussed in a subsequent section in Chapter 6.

Note that the weights generated in this section are fractional. For agent-based modelling applications integer weights are required, which can be generated through the process of integerisation. Two methods of integerisation are explored: The first randomly chooses the individuals, with a probability proportional to the weights. The second is to define a rounding method. A good method is 'Truncate Replicate Sample' (Lovelace and Ballas, 2013), described in (Section 5.3).

5.2.4 IPF with mipfp

The R package **mipfp** is a more generalized implementation of the IPF algorithm than **ipfp**, and was designed for population synthesis. **ipfp** generates a two-dimensional weight matrix based on mutually exclusive (non cross-tabulated) constraint tables. This is useful for applications using constraints, which are only marginals and which are not cross-tabulated. **mipfp** is more flexible, allowing multiple cross-tabulations in the constraint variables, such as age/sex and age/class combinations.

mipfp is therefore a multidimensional implementation of IPF, which can update an initial N-dimensional array with respect to given target marginal distributions. These, in turn, can also be multidimensional. In this sense **mipfp** is more advanced than **ipfp** which solves only the 2-dimensional case.

The main function of **mipfp** is `Ipfp()`, which fills a N-dimensional weight matrix based on a range of aggregate level constraint table options. Let's test the package on some example data. The first step is to load the package into the workspace:

```
library(mipfp) # after install.packages("mipfp")
```

To illustrate the use of `Ipfp`, we will create a fictive example. The case of SimpleWorld is here too simple to really illustrate the power of this package. The solving of SimpleWorld with `Ipfp` is included in the next section containing the comparison between the two packages.

The example case is as follows: to determine the contingency table of a population characterized by categorical variables for age (0-17, Workage, 50+), gender (male, female) and educational level (level 1 to level 4). We consider a zone with 50 inhabitants. The classic spatial microsimulation problem consists in having all marginal distributions and the cross-tabulated result (age/gender/education in this case) only for a non-geographical sample.

We consider the variables in the following order: sex (1), age (2) and diploma (3). Our constraints could for example be:

```
sex <- c(Male = 23, Female = 27) # n. in each sex category

age <- c(Less18 = 16, Workage = 20, Senior = 14) # age bands

diploma <- c(Level1 = 20, Level2 = 18, Level3 = 6, Level4 = 6)
```

Note the population is equal in each constraint (50 people). Note also the order in which each category is encoded — a common source of error in population synthesis. For this reason we have labelled each category of each constraint. The constraints are the target margins and need to be stored as a list. To tell the algorithm which elements of the list correspond to which constraint, a second list with the description of the target must be created. We print the *target* before running the algorithm to check the inputs make sense.

```
target <- list (sex, age, diploma)
descript <- list (1, 2, 3)
target
```

```
## [[1]]
##    Male Female
##      23     27
##
## [[2]]
##  Less18 Workage  Senior
##      16      20      14
##
## [[3]]
## Level1 Level2 Level3 Level4
##     20     18      6      6
```

Now that all constraint variables have been encoded, let us define the initial array to be updated, also referred to as the seed or the weight matrix. The dimension of this matrix must be identical to that of the constraint tables: $(2 \times 3 \times 4)$. Each cell of the array represents a combination of the attributes' values, and thus defines a particular category of individuals. In our case, we will consider that the weight matrix contains 0 when the category is not possible and 1 otherwise. In this example we will assume that it is impossible for an individual being less than 18 years old to hold a diploma level higher than 2. The corresponding cells are then set to 0, while the cells of the feasible categories are set to 1.

```
names <- list (names(sex), names(age), names(diploma))
weight_init <- array (1, c(2,3,4), dimnames = names)
weight_init[, c("Less18"), c("Level3","Level4")] <- 0
```

Everything being well defined, we can execute *Ipfp*. As with the package **ipfp**, we can choose a stopping criterion defined by *tol* and/or a maximum number of iterations. Finally, setting the argument `print` to `TRUE` ensures we will see the evolution of the procedure. After reaching either the maximum number of iterations or convergence (whichever comes first), the function will return a list containing the updated array, as well as other information about the convergence of the algorithm. Note that if the target margins are not consistent, the input data is then normalised by considering probabilities instead of frequencies.

```
result <- Ipfp(weight_init, descript, target, iter = 50,
               print = TRUE, tol = 1e-5)
```

```
## Margins consistency checked!
## ... ITER 1
##          stoping criterion: 4.236364
## ... ITER 2
##          stoping criterion: 0.5760054
## ... ITER 3
##          stoping criterion: 0.09593637
## ... ITER 4
##          stoping criterion: 0.01451327
## ... ITER 5
##          stoping criterion: 0.002162539
## ... ITER 6
##          stoping criterion: 0.0003214957
## ... ITER 7
##          stoping criterion: 4.777928e-05
## ... ITER 8
## Convergence reached after 8 iterations!
```

Note that the fit improves rapidly and attains the *tol* after 8 iterations. The `result` contains the final weight matrix and some information about the convergence. We can verify that the resulting table (`result$x.hat`) represents 50 inhabitants in the zone. Thanks to the definitions of names in the array, we can easily interpret the result: nobody younger than 18 has an educational level of 3 or 4. This is as expected.

```
result$x.hat # print the result
```

```
## , , Level1
##
##             Less18  Workage   Senior
## Male    3.873685 3.133126 2.193188
## Female 4.547370 3.678018 2.574612
##
## , , Level2
##
##             Less18  Workage   Senior
## Male    3.486317 2.819814 1.973870
## Female 4.092633 3.310216 2.317151
##
## , , Level3
##
##          Less18  Workage   Senior
## Male          0 1.623529 1.136471
## Female        0 1.905882 1.334118
##
## , , Level4
##
##          Less18  Workage   Senior
## Male          0 1.623529 1.136471
## Female        0 1.905882 1.334118
```

```
sum(result$x.hat) # check the total number of persons
```

```
## [1] 50
```

The quality of the margins with each constraints is contained in the variable
`check.margins` of the resulting list. In our case, we fit all constraints.

```
# printing the resulting margins
result$check.margins
```

```
## [1] 0.000000e+00 4.360648e-06 3.552714e-15
```

This reasoning works zone per zone and we can generate a 3-dimensional
weight matrix. Another advantage of **mipfp** is that it allows cross-tabulated
constraints to be added. In our example, we could add as target the contingency
of the age and the diploma. We can define this cross table:

```
# define the cross table
cross <- cbind(c(11,5,0,0), c(3,9,4,4), c(6,4,2,2))
rownames (cross) <- names (diploma)
colnames (cross) <- names(age)

# print the cross table
cross
```

```
##          Less18 Workage Senior
## Level1       11       3      6
## Level2        5       9      4
## Level3        0       4      2
## Level4        0       4      2
```

When having several constraints concerning the same variable, we have to ensure the consistency across these targets (otherwise convergence might not be reached). For instance, we can display the margins of `cross` along with `diploma` and `age` to validate the consistency assumption. There are both equal, meaning that the constraints are coherent.

```
# check pertinence for diploma
rowSums(cross)
```

```
## Level1 Level2 Level3 Level4
##     20     18      6      6
```

```
diploma
```

```
## Level1 Level2 Level3 Level4
##     20     18      6      6
```

```
# check pertinence for age
colSums(cross)
```

```
##  Less18 Workage  Senior
##      16      20      14
```

```
age
```

```
##  Less18 Workage  Senior
##      16      20      14
```

The `target` and `descript` have to be updated to include the cross table. Pay attention to the order of the arguments by declaring a contingency table. Then, we can execute the task and run *Ipfp* again:

```
# defining the new target and associated descript
target <- list(sex, age, diploma, cross)
descript <- list(1, 2, 3, c(3,2))

# running the Ipfp function
result <- Ipfp(weight_init, descript, target, iter = 50,
               print = TRUE, tol = 1e-5)
```

```
## Margins consistency checked!
## ... ITER 1
##        stoping criterion: 4.94
## ... ITER 2
## Convergence reached after 2 iterations!
```

The addition of a constraint leaves less freedom to the algorithm, implying that the algorithm converges faster. By printing the results, we see that all constraints are respected and we have a 3-dimensional contingency table.

Note that the steps of integerisation and expansion are pretty similar to the ones in **ipfp**. In the logic, it is exactly the same. The only difference is that for **ipfp** we consider vectors and for **mipfp** we consider arrays. For this reason, another function of expansion is created for arrays.

Finally it should also be noted that the **mipfp** package provides other methods to solve the problem such as the maximum likelihood, minimum chi-square and least squares approaches along with functionnalities to assess the accuracy of the produced arrays.

5.3   Integerisation

5.3.1   Concept of integerisation

The weights generated in the previous section are fractional, making the results difficult to use as a final table of individuals, needed as input to agent-based models. *Integerisation* refers to methods for converting these fractional weights into integers, with a minimum loss of information (Lovelace and Ballas, 2013). Simply rounding the weights is one integerisation method, but the results are very poor. This process can also be referred to as 'weighted sampling without replacement', as illustrated by the R package wrswoR (`https://github.com/krlmlr/wrswoR`).

Two integerisation methods are tackled in this chapter. The first treats weights as simple probabilities. The second constrains the maximum and minimum integer weight that can result from the integer just above and just under each fractional weight, and is known as TRS: 'Truncate Replicate Sample'.

Integerisation is the process by which a vector of real numbers is converted into a vector of integers corresponding to the number of individuals present in synthetic spatial microdata. The length of the new vector must equal the weight vector and high weights must be still high in the integerised version. Thus, high weights must be sampled proportionally more frequently than those with low weights for the operation to be effective. The following example illustrates how the process, when seen as a function called *int*, would work on a vector of 3 weights:

$$w_1 = (0.333, 0.667, 3)$$

$$int(w_1) = (0, 1, 3)$$

Note that $int(w_1)$ is a vector of same length. Then, $int(w_1)$ will be the result of the integerisation and each cell will contain an integer corresponding to the number of replication of this individual needed for this zone. This result was obtained by calculating the sum of the weights (4, which represents the total population of the zone) and sampling from these until the total population is reached. In this case individual 2 is selected once as they have a weight approaching 1, individual 3 will have to be replicated (*cloned*) three times and individual 1 does not appear in the integerised dataset at all, as it has a low weight. In this case the outcome is straightforward because the numbers are small and simple. But what about in less clear-cut cases, such as $w_2 = (1.333, 1.333, 1.333)$? What is needed is an algorithm to undertake this process of *integerisation* systematically to maximize the fit between the synthetic and constraint data for each zone.

There are a number of integerisation strategies available. Lovelace and Ballas (2013) tested 5 of these and found that the probabilistic methods of *proportional probabilities* and *TRS* outperformed their deterministic rivals. The first of these integerisation strategies can be written as an R functions as follows:

```
int_pp <- function(x){
  # For generalisation purpose, x becomes a vector
  # This allow the function to work with matrix also
  xv <- as.vector(x)
```

```
# Initialisation of the resulting vector
xint <- rep(0, length(x))
# Sample the individuals
rsum <- round(sum(x))
xs <- sample(length(x), size = rsum, prob = x, replace = T)
# Aggregate the result into a weight vector
xsumm <- summary(as.factor(xs))
# To deal with empty category, we restructure
topup <- as.numeric(names(xsumm))
xint[topup] <- xsumm
# For generalisation purpose, x becomes a matrix if necessary
dim(xint) <- dim(x)
xint
}
```

The R function `sample` needs in the order the arguments: the set of objects that can be chosen, the number of randomly drawn and the probability of each object to be chosen. Then we can add the argument `replace` to tell R if the sampling has to be with replacement or not. To test this function let's try it on the vectors of length 3 described in code:

```
set.seed(50)
xint1 <- int_pp(x = c(0.333, 0.667, 3))
xint1
```

```
## [1] 0 1 3
```

```
xint2 <- int_pp(x = c(1.333, 1.333, 1.333))
xint2
```

```
## [1] 1 1 2
```

Note that it is probabilistic, so normally if we proceed two times the same, the results can be different. So, by fixing the `seed`, we are sure to obtain each time the same result. The result here was the same as that obtained through intuition. However, we can re-execute the same code and obtain different results:

```
xint1 <- int_pp(x = c(0.333, 0.667, 3))
xint1
```

```
## [1] 0 0 4
```

```
xint2 <- int_pp(x = c(1.333, 1.333, 1.333))
xint2
```

```
## [1] 1 2 1
```

The result for `xint1` is now not so intuitive: one would expect having 3 times the third individual and 1 time the first or second one. But, you have to be aware that a random choice means that even the cell with low probability can be chosen, but not so many times. This also implies the possibility to choose an individual more times than intuitively though, since the weights are not adapted during the sampling. Here, we draw only very few numbers, thus it is normal that the resulting distribution is not exactly the same as the wanted. However, probabilities mean that if you have a very large dataset, your random choice will result in a distribution very close to the desired one. We can generate 100,000 integers between 1 and 100,000 and see the distribution. Figure 5.3 contains the associated histogram and we can see that each interval occurs more or less the same number of times.

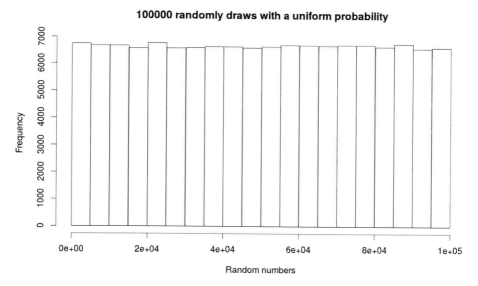

FIGURE 5.3
Distribution of a vector generated randomly (each integer between 1 and 100,000 being equally likely).

An issue with the *proportional probabilities* (PP) strategy is that completely unrepresentative combinations of individuals have a non-zero probability of being sampled. The method will output $(4, 0, 0)$ once in every 21 thousand runs for w_1 and once every 81 runs for w_2. The same probability is allocated to all other 81 (3^4) permutations.[6]

To overcome this issue Lovelace and Ballas (2013) developed a method which ensures that any individual with a weight above 1 would be sampled at least once, making the result $(4, 0, 0)$ impossible in both cases. This method is *truncate, replicate, sample* (TRS) integerisation, described in the following function:

```
int_trs <- function(x){
  # For generalisation purpose, x becomes a vector
  xv <- as.vector(x) # allows trs to work on matrices
  xint <- floor(xv) # integer part of the weight
  r <- xv - xint # decimal part of the weight
  def <- round(sum(r)) # the deficit population
  # the weights be 'topped up' (+ 1 applied)
  topup <- sample(length(x), size = def, prob = r)
  xint[topup] <- xint[topup] + 1
  dim(xint) <- dim(x)
  dimnames(xint) <- dimnames(x)
  xint
}
```

This method consists of 3 steps : truncate all weights by keeping only the integer part and forget the decimals. Then, replicate by considering these integers as the number of each type of individuals in the zone. And finally, sample to reach the good number of people in the zone. This stage is using a sampling with the probability corresponding to the decimal weights. To see how this new integerisation method and associated R function performed, we run it on the same input vectors:

```
set.seed(50)
xint1 <- int_trs(x = c(0.333, 0.667, 3))
xint1
```

```
## [1] 1 0 3
```

```
xint2 <- int_trs(x = c(1.333, 1.333, 1.333))
xint2
```

---

[6]There are 3^4 possibilities because 3 options are allocated to 4 'spaces', due to the total population being 4. If the total population in this example were 2, there would be 9 ways of allocating the indices 1, 2 and 3: there are 9 *permutations*. The general rule governing the number of permutations is $*p = n^r$ where n is the number of options and r is the number of repetitions.

```
## [1] 1 1 2
```

```
# Second time:
xint1 <- int_trs(x = c(0.333, 0.667, 3))
xint1
```

```
## [1] 0 1 3
```

```
xint2 <- int_trs(x = c(1.333, 1.333, 1.333))
xint2
```

```
## [1] 2 1 1
```

Note that while `int_pp` (the *proportional probabilities* function) produced an output with 4 instances of individual 3 (in other words allocated individual 3 a weight of 4 for this zone), TRS allocated a weight 1 either for the first or for the second individual. Although the individual that is allocated a weight of 2 will vary depending on the random seed used in TRS, we can be sure that TRS will never convert a fractional weight above one into an integer weight of less than one. In other words the range of possible integer weight outcomes is smaller using TRS instead of the PP technique. TRS is a more tightly constrained method of integerisation. This explains why spatial microdata produced using TRS fit more closely to the aggregate constraints than spatial microdata produced using different integerisation algorithms (Lovelace and Ballas 2013). This is why we use the TRS methodology, implemented through the function `int_trs`, for integerisation throughout the majority of this book.

5.3.2 Example of integerisation

Let's use TRS to generate spatial microdata for SimpleWorld. Remember, we already have generated the weight matrix `weights`. To integerise the weights of first zone, we just need to code:

```
int_weight1 <- int_trs(weights[,1])
```

To analyse the result, we just print the weights for zone 1 before and after integerisation. This shows a really intuitive result.

```
weights[,1]
```

```
## [1] 1.227998 1.227998 3.544004 1.544004 4.455996
```

```
int_weight1
```

```
## [1] 1 1 4 2 4
```

TRS worked in this case by selecting individuals 1 and 4 and then selecting two more individuals (2 and 4) based on their non-integer probability after truncation.

When using the **mipfp** package, the result is a cross table per category instead of a weight vector per individual. Consequently, we must repeat the categories instead of the individual. However, the integerisation step is exactly similar, since we just need to have integers to replace the doubles. The function `int_trs` presented before is applicable on this 3-dimensional array.

```
set.seed(42)
mipfp_int <- int_trs(result$x.hat)
```

The variable `mipfp_int` contains the integerised weights. This means that the cross table of all target variables is stored in `mipfp_int`.

We can look at the table before and after integerisation. To be more concise, we just print the sub-table corresponding to the first level of education.

```
# Printing weights for first level of education
# before integerisation
result$x.hat[,,1]
```

```
##          Less18 Workage Senior
## Male       5.06    1.38   2.76
## Female     5.94    1.62   3.24
```

```
# Printing weights for first level of education
# after integerisation
mipfp_int[,,1]
```

```
##          Less18 Workage Senior
## Male          5       2      3
## Female        6       2      4
```

The integerisation has just chosen integers close to the weight. Note that this step can introduce some error in the marginal totals. However, the maximal error is always 1 per category, which is usually small compared with the total population. We could develop an algorithm to ensure perfect fit after integerisation this would add much complexity and computational time version for little benefit. To generate the final microdata, the final step is expansion.

5.4 Expansion

As illustrated in Figure 5.1, the remaining step to create the spatial microdata is expansion. Depending on the chosen method, you can have as result either a vector of weights per individual (for example with **ipfp**), or a matrix of weights per different possible category (for example with **mipfp**). The expansion step will be a bit different if you have one or the other structure of weights.

5.4.1 Weights per individual

In this case each integerised weight corresponds to the number of needed repetitions for each individual. For this purpose, we will first generate a vector of IDs of the sampled individuals. This can be achieved with the function `int_expand_vector`, defined below:

```
int_expand_vector <- function(x){
  index <- 1:length(x)
  rep(index, round(x))
}
```

This function returns a vector. Each element of this vector represents an individual in the final spatial microdata and each cell will contain the ID of the associated sampled individual. This is illustrated in the code below, which uses the integerised weights for zone 1 of SimpleWorld, generated above.

```
int_weight1 # print weights
```

```
## [1] 1 1 4 2 4
```

```
# Expand the individual according to the weights
exp_indices <- int_expand_vector(int_weight1)
exp_indices
```

```
## [1] 1 2 3 3 3 3 4 4 5 5 5 5
```

The third weight was 4 and the associated `exp_indices` contains 4 times the ID '3'. The final data can be found simply by replicating the individuals.

```
# Generate spatial microdata for zone 1
ind_full[exp_indices,]
```

```
##      id age sex income
## 1    1  59   m   2868
## 2    2  54   m   2474
## 3    3  35   m   2231
## 3.1  3  35   m   2231
## 3.2  3  35   m   2231
## 3.3  3  35   m   2231
## 4    4  73   f   3152
## 4.1  4  73   f   3152
## 5    5  49   f   2473
## 5.1  5  49   f   2473
## 5.2  5  49   f   2473
## 5.3  5  49   f   2473
```

5.4.2   Weights per category

This stage is different from the expansion developed above because the structure of data is totally different. Indeed, above we simply need to replicate the individuals, whereas in this case, we must retrieve the category names to recreate the individuals.

This process is included inside the function `int_expand_array`. This little function first saves a data frame version of the integerised matrix. Then, the `count_data` contains a line for each different combination of categories (gender, sex, diploma) and the corresponding number of individuals. Having this, the remaining task is to consider only categories having more than 0 counts and repeat them the right number of times.

```
int_expand_array <- function(x){
  # Transform the array into a dataframe
  count_data <- as.data.frame.table(x)
  # Store the indices of categories for the final population
  indices <- rep(1:nrow(count_data), count_data$Freq)
  # Create the final individuals
  ind_data <- count_data[indices,]
  ind_data
}
```

We apply this function to our data and the result is a `data.frame` object, whose rows represent individuals. The first column is the gender, the second is age and the third is the education level. We print the head of this data frame and we can check that the individual of type (Male, Less18, Level1) is repeated 5 times, which is its weight.

```
# Expansion step
ind_mipfp <- int_expand_array(mipfp_int)

# Printing final results
head(ind_mipfp)
```

```
##          Var1    Var2    Var3 Freq
## 1        Male Less18 Level1    5
## 1.1      Male Less18 Level1    5
## 1.2      Male Less18 Level1    5
## 1.3      Male Less18 Level1    5
## 1.4      Male Less18 Level1    5
## 2      Female Less18 Level1    6
```

Note that this example was designed for only one zone. A similar reasoning is applicable to as many zones as you want.

5.5 Integerisation and expansion

The process to transform the weights into spatial microdata is now performed for all zones. This section demonstrates two ways to generate spatial microdata.[7]

```
# Method 1: using a for loop
ints_df <- NULL
set.seed(42)
for(i in 1:nrow(cons)){
  # Integerise and expand
  ints <- int_expand_vector(int_trs(weights[, i]))
  # Take the right individuals
  data_frame <- data.frame(ind_full[ints,])
  ints_df <- rbind(ints_df, data_frame, zone = i)
}
```

---

[7]Note that if you skipped the detailed previous sections, you will need to load the functions by executing `source("R/functions.R")`.

```
# Method 2: using apply
set.seed(42)
ints_df2 <- NULL
# Take the right indices
ints2 <- unlist(apply(weights, 2, function(x)
  int_expand_vector(int_trs(x))))
# Generate the individuals
ints_df2 <- data.frame(ind_full[ints2,],
  zone = rep(1:nrow(cons), colSums(weights)))
```

Both methods yield the same result for `ints_df`, the only differences being that Method 1 is perhaps more explicit and easier to understand whilst Method 2 is more concise.

`ints_df` represents the final spatial microdataset, representing the entirety of SimpleWorld's population of 33 (this can be confirmed with `nrow(ints_df)` individuals). To select individuals from one zone only is simple using R's subsetting notation. To select all individuals generated for zone 2, for example, the following code is used. Note that this is the same as the output generated in Table 5 at the end of the SimpleWorld chapter — we have successfully modelled the inhabitants of a fictional planet, including income!

```
ints_df[ints_df$zone == 2, ]
```

```
## [1] id      age     sex     income
## <0 rows> (or 0-length row.names)
```

Note that we take here the example for the case of weights per individual, but the process for weights per category is exactly the same except that you need to use the function `int_expand_array` instead of `int_expand_vector` and checking that the `i` is in the dimension corresponding to the zone. In the previous example for the array, we didn't execute the algorithm for several zones. However, this process will be explained in the comparison section.

5.6   Comparing ipfp with mipfp

Although designed to handle large datasets, **mipfp** can also solve smaller problems such as SimpleWorld. We have simply chosen to develop here a different example to show you the vast application of this function. However, to compare **ipfp** with **mipfp**, we will focus on this simpler example: SimpleWorld.

5.6.1   Comparing the methods

A major difference between the packages **ipfp** and **mipfp** is that the former is written in C, whilst the latter is written in R. Another is that the *aim* of **mipfp** is to execute IPF for spatial microsimulation, accepting a wide variety of inputs (cross-table or marginal distributions). **ipfp**, by contrast, was created to solve an algebra problem of the form $Ax = b$. Its aim is to find a vector x such as $Ax = b$, after having predefined the vector b and the matrix A. Thus, the dimensions are fixed : a two-dimensional matrix and two vectors. In addition, **mipfp** is written in R, so should be easy to understand for R users. **ipfp**, by contrast, is written in pure C: fast but relatively difficult to understand.

As illustrated earlier in the book, **ipfp** gives weights to every individual in the input microdata, for every zone. **mipfp** works differently. Individuals who have the same characteristics (in terms of the constraint variables) appear only once in the input 'seed', their number represented by their starting weight. **mipfp** thus determines the number of people in each unique combination of constraint variable categories. This makes **mipfp** more *computationally efficient* than **ipfp**. This efficiency can be explained for SimpleWorld. We have constraints about sex and age. What will be defined is the number of persons in each category so that the contingency table respects the constraints. In other words, we want, for zone 1 of SimpleWorld, to fill in the unknowns in Table 5.1.

sex - age	0-49 yrs	50 + yrs	Total
m	?	?	6
f	?	?	6
Total	8	4	12

TABLE 5.1: Unknown constraints Mipfp estimates for SimpleWorld.

From this contingency table, and respecting the known marginal constraints, the individual dataset must be generated. For SimpleWorld, input microdata is available. While **ipfp** uses the individual level microdata directly, the 'seed' used in **mipfp** must also be a contingency table. This table is then used to generate the final contingency table (sex/age) thanks to the theory underlying IPF.

For the first step, we begin with the aggregation of the microdata. We have initially a total of 5 individuals and we want to have 12. During the steps of IPF, we will always consider the current totals and desired totals (Table 5.2).

sex - age	0-49 yrs	50 + yrs	Total	Target total
m	1	2	3	6
f	1	1	2	6
Total	2	3		
Target total	8	4		

TABLE 5.2: Initial table of Mipfp for SimpleWorld.

IPF makes a proportion of the existing table to fit each constraint turn by turn. First, if we want to fit the age constraint, we need to sum 8 persons under 50 years old and 4 above (wanted total of age). Thus, we calculate a proportional weight matrix that respects this constraint. The first cell corresponds to $Current_{Weight} * \frac{Wanted_{Total}}{Current_{Total}} = 1 * \frac{8}{2}$. With a similar reasoning, we can complete the whole matrix (Table 5.3).

sex - age	0-49 yrs	50 + yrs	Total	Wanted total
m	4	2.7	6.7	6
f	4	1.3	5.3	6
Total	8	4		
Wanted total	8	4		

TABLE 5.3: First step of Mipfp for SimpleWorld - age constraint.

Now, the current matrix fits exactly the age constraint — note how this was achieved ($1 * \frac{8}{2}$ for the first cell). However, there are still differences between the current and target sex totals. IPF proceeds to make the table fit for the sex constraint (Table 5.4).

sex - age	0-49 yrs	50 + yrs	Total	Wanted total
m	3.6	2.4	6	6
f	4.5	1.5	6	6
Total	8.1	3.9		
Wanted total	8	4		

TABLE 5.4: First step of Mipfp for SimpleWorld - sex constraint.

This ends the first iteration of ipfp, illustrated in the form of contingency table. **mipfp** simply automates the logic, but by watching populations in terms of tables. When the contingency table is complete, we know how many individuals we need in each category. As with the **ipfp** method, the weights are fractional. To reach a final individual level dataset, we can consider the weights as probabilities or transform the weight into integers and obtain a final contingency table. From this, it is easy to generate the individual dataset.

It should be clear from the above that both methods perform the same procedure. The purpose of this section, beyond demonstrating the equivalence of **ipfp** and **mipfp** approaches to population synthesis, is to compare the *performance* of each approach. Using the simple example of SimpleWorld we first compare the results before assessing the time taken by each package.

5.6.2   Comparing the weights for SimpleWorld

The whole process using **ipfp** has been developed in the previous section. Here, we describe how to solve the same SimpleWorld problem with **mipfp**. We proceed zone-by-zone, explaining in detail the process for zone 1. The constraints for this area are as follows:

```
# SimpleWorld constraints for zone 1
con_age[1,]
```

```
##   a0_49 a50+
## 1     8    4
```

```
con_sex[1,]
```

```
##   m f
## 1 6 6
```

```
# Save the constraints for zone 1
con_age_1 <- as.matrix( con_age[1,] )
con_sex_2 <- data.matrix( con_sex[1, c(2,1)] )
```

The inputs of `Ipfp` must be a list. We consider the age first and then the sex. Note that the order of the sex categories must be the same as those of the seed. Checking the order of the category is sensible as incorrect ordering here is a common source of error. If the seed has (Male, Female) and the constraint (Female, Male), the results will be wrong.

```
# Save the target and description for Ipfp
target <- list (con_age_1, con_sex_2)
descript <- list (1, 2)
```

The available sample establishes the seed of the algorithm. This is also known as the *initial weight matrix*. This represents the entire individual level input microdata (`ind`), converted into the same categories as in the constraints (50 is still the break point for ages in SimpleWorld). We also consider income, so that it is stored in the format required by **mipfp**. (An alternative would be to ignore income here and add it after, as seen in the expansion section.)

```
# load full ind data
ind_full <- read.csv("data/SimpleWorld/ind-full.csv")
# Add income to the categorised individual data
ind_income <- cbind(ind, ind_full[,c("income")])
weight_init <- table(ind_income[, c(2,3,4)])
```

Having the initial weight initialised to the aggregation of the individual level data and having the constraints well established, we can now execute the `Ipfp` function.

```
# Perform Ipfp on SimpleWorld example
weight_mipfp <- Ipfp(weight_init, descript, target,
  print = TRUE, tol = 1e-5)
```

```
## Margins consistency checked!
## ... ITER 1
##       stoping criterion: 3.5
## ... ITER 2
##       stoping criterion: 0.05454545
## ... ITER 3
##       stoping criterion: 0.001413095
## ... ITER 4
##       stoping criterion: 3.672544e-05
## ... ITER 5
## Convergence reached after 5 iterations!
```

Convergence was achieved after only 5 iterations. The result, stored in the `weight_mipfp` is a `data.frame` containing the final weight matrix, as well as additional outputs from the algorithm. We can check the margins and print the final weights. Note that we still have the income into the matrix. For the comparison, we look only at the result for the aggregate table of age and sex.

```
# Printing the resulting margins
weight_mipfp$check.margins
```

```
## [1] 4.560676e-08 0.000000e+00
```

```
# Printing the resulting weight matrix for age and sex
apply(weight_mipfp$x.hat, c(1,2), sum)
```

```
##          sex
## age              f         m
##    a0_49 4.455996 3.544004
##    a50+  1.544004 2.455996
```

We can see that we reach the convergence after 5 iterations and that the results are fractional, as with **ipfp**. Now that the process is understood for one zone, we can progress to generating weights for each zone. The aim is to transform the output into a form similar to the one generated in the section **ipfp** to allow an easy comparison. For each area, the initial weight matrix will be the same. We store all weights in an array `Mipfp_Tab`, in which each row represents a zone.

```
# Initialising the result matrix
Names <- list(1:3, colnames(con_sex)[c(2,1)], colnames(con_age))
Mipfp_Tab <- array(data = NA, dim = c(n_zone, 2 , 2),
                   dimnames = Names)

# Loop over the zones and execute Ipfp
for (zone in 1:n_zone){
  # Adapt the constraint to the zone
  con_age_1 <- data.matrix(con_age[zone,] )
  con_sex_2 <- data.matrix(con_sex[zone, c(2,1)] )
  target <- list(con_age_1, con_sex_2)

  # Calculate the weights
  res <- Ipfp(weight_init, descript, target, tol = 1e-5)

  # Complete the array of calculated weights
  Mipfp_Tab[zone,,] <- apply(res$x.hat,c(1,2),sum)
}
```

The above code executes IPF for each zone. We can print the result for zone 1:2 and observe that it corresponds to the weights previously determined.

```
# Result for zone 1
Mipfp_Tab[1:2,,]
```

```
##  , , a0_49
##
##            f         m
## 1 4.455996 1.544004
## 2 1.450166 4.549834
##
##  , , a50+
##
##            f         m
## 1 3.5440038 2.455996
## 2 0.5498344 3.450166
```

Cross-tabulated constraints can be used with **mipfp**. This means that we can consider the zone as a variable, allowing population synthesis to be performed with no `for` loop, to generate a result for every zone. This could seem analogous to the use of `apply()` in conjunction with `ipfp()`, but it is not really the case. Indeed, the version with `apply()` calls `n_zone` times the function `ipfp()`, where the **mipfp** is called only one time with the here proposed abbreviation.

Only one aspect of our model setup needs to be modified: the number of the zone needs to be included as an additional dimension in the weight matrix. This allows us to access the table for one particular zone and obtain exactly the same result as previously determined.

Thus, as before, we first need to define an initial weight matrix. For this, we repeat the current initial data as many times as the number of zones. The variables are created in the following order: zone, age, sex, income. Note that zone is the first dimension here, but it could be be the second, third or last.

```
# Repeat the initial matrix n_zone times
init_cells <- rep(weight_init, each = n_zone)

# Define the names
names <- c(list(c(1, 2, 3)), as.list(dimnames(weight_init)))

# Structure the data
mipfp_zones <- array(init_cells,
  dim = c(n_zone, n_age, n_sex, 5),
  dimnames = names)
```

The resulting array is larger than the ones obtained with `ipfp()`, containing `n_zone` times more cells. Instead of having `n_zone` different little matrices, the result is here a bigger matrix. In total, the same number of cells are recorded. To access all information about the first zone, type `mipfp_zones[1,,]`. Let's check that this corresponds to the initial weights used in the `for` loop example:

```
table(mipfp_zones[1,,,] == weight_init)
```

```
##
## TRUE
##   20
```

The results show that the 20 cells of zone 1 are equal in value to those in the previous matrix. With these initial weights, we can keep the whole age and sex constraint tables (after checking the order of the categories). The constraint of age is in this case cross-tabulated with zone, with the first entry the zone and second the age.

```
con_age
```

```
##   a0_49 a50+
## 1     8    4
## 2     2    8
## 3     7    4
```

Next the `target` and the `descript` variables must be adapted for use without a `for` loop. The new target takes the two whole matrices and the `descript` associates the first constraint with (zone, age) and the second with (zone, sex).

```
# Adapt target and descript
target <- list(data.matrix(con_age),
  data.matrix(con_sex[,c(2,1)]))
descript <- list (c(1,2), c(1,3))
```

We are now able to perform the entire execution of `Ipfp`, for all zones, in only one line.

```
res <- Ipfp(mipfp_zones, descript, target, tol = 1e-5)
```

This process gives exactly the same result as the for loop but is more practical and efficient. For all applications, the 'no for loop' strategy is preferred, being faster and more concise. However, in this section, we aim to compare the two processes for specific zones. For this reason, to make things comparable for little area, we will keep the 'for loop', allowing an equitable comparison of time for one iteration.

Note that the output of the function `Ipfp` can be used directly to perform steps such as integerisation and expansion. However, here we first verify that both algorithms calculate the same weights. For this we must transform the output of `Ipfp` from **mipfp** to match the weight matrix generated by `ipfp` from the **ipfp** package. The comparison table will be called `weights_mipfp`. To complete its cells, we need the number of individuals in the sample in each category.

```
# Initialise the matrix
weights_mipfp <- matrix(nrow = nrow(ind), ncol = nrow(cons))

# Cross-tabulated contingency table of the microdata, ind
Ind_Tab <- table(ind[,c(2,3)])
```

Now, we can fill this matrix `weights_mipfp`. For each zone, the weight of the category is distributed between the individuals of the microdata having the right characteristics. This is done thanks to iterations that calculate the weights per category for the zone and then transform them into weights per individual, depending on the category of each person.

```
# Loop over the zones to transform the result's structure
for (zone in 1:n_zone){

  # Transformation into individual weights
  for (i in 1:n_ind){
    # weight of the category
    weight_ind <- Mipfp_Tab[zone, ind[i, 2], ind[i, 3]]

    # number of ind in the category
    sample_ind <- Ind_Tab[ind[i, 2], ind[i, 3]]

    # distribute the weight to the ind of this category
    weights_mipfp[i,zone] <- weight_ind / sample_ind
  }
}
```

The results from **mipfp** are now comparable to those from **ipfp**, since we generated weights for each individual in each zone. Having the same structure, the integerisation and expansion step explained for **ipfp** could be followed. The `int_expand_array()` provides an easier way to expand these weights (see section 5.4.2).

The below code demonstrates that the largest difference is of the order 10^{-7}, which is negligible. Thus, we have demonstrated that both functions generate the same result.

```
# Difference between weights of ipfp and mipfp
abs(weights - weights_mipfp)
```

```
##            [,1]          [,2]          [,3]
## [1,] 1.273616e-08 2.868000e-09 2.207059e-08
## [2,] 1.273616e-08 2.868000e-09 2.207059e-08
## [3,] 2.547231e-08 5.736000e-09 4.414117e-08
## [4,] 2.013445e-08 1.330023e-08 1.023514e-07
## [5,] 2.013445e-08 1.330023e-08 1.023514e-07
```

One package is written in C and the other in R. The current question is : Which package is more efficient? The answer will depend on your application. We have seen that both algorithms do not consider the same form of input. For **ipfp** you need constraints and an initial individual level data, coming for example from a survey. For **mipfp** the constraints and a contingency table of a survey is enough. Thus, in the absence of microdata, **mipfp** must be used (described in Chapter 9). Moreover, their results are in different structures. Weights are created for each individual with **ipfp** and weights for the contingency table with **mipfp**. The results can be transformed from one to the other form if needs be.

5.6.3 Comparing the results for SimpleWorld

In this section, we compare the final dataset, after integerisation and expansion, for SimpleWorld. The final individuals are stored in the variable `ints_df` for the package **ipfp**.

On the other side, for **mipfp**, the integerisation and expansion steps need to be done. First, note that it is better to integerise zone per zone. Indeed, the method consists in drawing randomly the remaining individuals. However, if we execute it for all zones together, it is possible that an individual will be drawn in a zone already full and another zone will be under-estimated.

```
# Save the matrix in a new variable
int_mipfp <- res$x.hat
# Integerise zone per zone
for (i in 1:n_zone){
  int_mipfp[i,,] <- int_trs(int_mipfp[i,,,])
}
```

Now, we can expand the result by using the same function as previously and call the final microdata `indiv_mipfp`.

```
# Expansion of mipfp integerised weight matrix
indiv_mipfp <- int_expand_array(int_mipfp)
Names <- c("Zone", "Age", "Sex", "Income", "Freq")
colnames(indiv_mipfp) <- Names
```

We print the result of both methods for zone 2. We observe that the **ipfp**
package gives the precise age, whereas **mipfp** includes the categories. However,
it is sometimes better to stay in the category level, since the ages possible are
only the ones present in the sample data. Moreover, due to the integerisation
step, the constraints are not perfectly fitted. Indeed, the **ipfp** result should
include a supplementary man and **mipfp** a young person.

```
ints_df[ints_df$zone == 2, ]
```

```
## [1] id       age      sex      income
## <0 rows> (or 0-length row.names)
```

```
indiv_mipfp[indiv_mipfp$Zone==2,]
```

```
##          Zone   Age Sex Income Freq
## 14          2 a0_49   f   2473    2
## 14.1        2 a0_49   f   2473    2
## 35          2 a50+    m   2474    1
## 47          2 a50+    m   2868    2
## 47.1        2 a50+    m   2868    2
## 53          2 a50+    f   3152    5
## 53.1        2 a50+    f   3152    5
## 53.2        2 a50+    f   3152    5
## 53.3        2 a50+    f   3152    5
## 53.4        2 a50+    f   3152    5
```

Usually, when dealing with big problems, the relative error caused by this step
is small.

5.6.4   Speed comparisons

This section compares the computational time of **mipfp** and **ipfp** approaches.

```
ind_mipfp <- int_expand_array(int_trs(res$x.hat))
```

The above code measures the time taken to execute the `ipfp` function for the
zone 1. It takes 0.001 second on a modern computer.[8] The code below makes

---

[8]The tests were performed on a computer with Intel i7-4930K processor running at
3.40GHz.

a similar analysis for the `Ipfp` function and takes 0.002 second on the same computer. Thus, for this very little example, the package **ipfp** written in C is faster. The first lines of code below are defined to ensure that we perform the same calculation. Thus, the times are comparable.

```
# Choose correct constraint for mipfp
con_age_1 <- data.matrix( con_age[1,] )
con_sex_2 <- data.matrix( con_sex[1, c(2,1)] )
target <- list (con_age_1, con_sex_2)
descript <- list (1, 2)

# Equitable calculus - remove income
weight_init_no_income <- table( ind[, c(2,3)])

# Measure time mipfp
system.time(Ipfp( weight_init_no_income, descript, target,
              tol = 1e-5))
```

An advantage of the **mipfp** package, for problems with a lot of input individuals, is in terms of memory. Indeed, **ipfp** needs a table with as many rows as individuals, whereas **mipfp** needs a table of a dimension corresponding to the number of different categories. For SimpleWorld, the inputs of **ipfp** and **mipfp** are respectively:

```
# input for ipfp
ind_cat
```

```
##   ind$agea0_49 ind$agea50+ ind$sexm ind$sexf
## 1            0           1        1        0
## 2            0           1        1        0
## 3            1           0        1        0
## 4            0           1        0        1
## 5            1           0        0        1
```

```
# input for mipfp
Ind_Tab
```

```
##         sex
## age      f m
##   a0_49  1 1
##   a50+   1 2
```

Imagine that we have a similar problem, but with a total population of 2000 and 500 initial individuals. The dimension of `ind_cat` will become 500 rows for 4 columns, where the one of `Ind_Tab` will not change.

By testing the time for SimpleWorld when considering having 25,000 individuals (we replicate the data 5000 times) and generating 12,000,000 individuals (we multiply the constraints by 1000), we obtain on the same computer that **ipfp** and **mipfp** take, respectively, 0.012 and 0.005 seconds.

In conclusion to this comparison, **mipfp** is more adaptable than **ipfp**. The most appropriate solution may depend on the context, such as the structure of the input data. In terms of computational time, the comparison demonstrates that **mipfp** was slower for small problems and faster for bigger problems. In all cases, they calculate the same results. Importantly for agent-based modelling, the weight matrices generated by both methods must undergo processes of *expansion* and *integerisation* to be converted into spatial microdata.

5.7   Chapter summary

In this chapter we have first seen what are weighting algorithms and the similarities between these algorithms and combinatorial optimisation (even if the underlying method is completely different). Then, we explained in detail what Iterative Proportional Fitting is, how it works and several ways to code in R. We continued by explaining how to have more realistic weights : integers. After that, a section contained the process to transform the weights into a final table containing all generated individuals with their characteristics.

Finally, we compared two packages for IPF and found **mipfp** to be more generalisable and powerful. The choice of approach should be influenced by the size and nature of the input data.

6

Alternative approaches to population synthesis

CONTENTS

This chapter briefly describes other techniques for spatial microsimulation. We have different methods, each corresponding to a section:

- *GREGWT* (Section 6.1) contains a method based on a Generalized Regression Weighting procedure.
- *Population synthesis as an optimization problem* (Section 6.2) explains how we can make spatial microsimulation thanks to optimization algorithms.
- *simPop* (Section 6.3) mentions another package to make spatial microsimulation.
- *The Urban Data Science Toolkit (UDST)* (Section 6.4) shortly describes another technique, coded in python.

6.1 GREGWT

As described in the Introduction, IPF is just one strategy for obtaining a spatial microdataset. However, researchers tend to select one method that they are comfortable with and stick with that for their models. This is understandable because setting up the method is usually time consuming: most researchers rightly focus on applying the methods to the real world rather than fretting about the details. On the other hand, if alternative methods work better for a particular application, resistance to change can result in poor model fit. In the case of very large datasets, spatial microsimulation may not be possible unless certain methods, optimised to deal with large datasets, are used. Above all, there is no consensus about which methods are 'best' for different applications, so it is worth experimenting to identify which method is most suitable for each application.

An interesting alternative to IPF method is the GREGWT algorithm. First implemented in the SAS language by the Statistical Service area of the Australian Bureau of Statistics (ABS), the algorithm reweighs a set of initial weights using a Generalized Regression Weighting procedure (hence the name GREGWT). The resulting weights ensure that, when aggregated, the individuals selected for each small area fit the constraint variables. Like IPF, the GREGWT results in non-integer weights, meaning some kind of integerisation algorithm will be needed to obtain individual level microdata. For example, if the output is to be used in ABM, the macro developed by ABS adds a weight restriction in their GREGWT macros to ensure positive weights. The ABS uses the Linear Truncated Method described in Singh and Mohl (1996) to enforce these restrictions.

A simplified version of this algorithm (and other algorithms) is provided by Rahman (2009). The algorithm is described in more detail in Tanton et al. (2011). An R implementation of GREGWT can be found in the GitHub repository GREGWT (https://github.com/emunozh/GREGWT) and installed using the function install_github() from the **devtools** package.

The code below uses this implementation of the GREGWT algorithm, with the data from Chapter 3.

```
# Install GREGWT (uncomment/alter as appropriate)
# devtools::install_github("emunozh/GREGWT")
# load the library (0.7.0)
library('GREGWT')

# Load the data from csv files stored under data
age = read.csv("data/SimpleWorld/age.csv")
sex = read.csv("data/SimpleWorld/sex.csv")
ind = read.csv("data/SimpleWorld/ind-full.csv")
# Make categories for age
ind$age <- cut(ind$age, breaks=c(0, 49, Inf),
               labels = c("a0.49", "a.50."))
# Add initial weights to survey
ind$w <- vector(mode = "numeric", length=dim(ind)[1]) + 1

# prepare simulation data using GREGWT::prepareData
data_in <- prepareData(cbind(age, sex),
                   ind, census_area_id = F, breaks = c(3))
```

```
# prepare a data.frame to store the result
fweights <- NULL
Result <- as.data.frame(matrix(NA, ncol=3, nrow=dim(age)[1]))
names(Result) <- c("area", "income", "cap.income")
```

The code presented above loads the SimpleWorld data and creates a new
data.frame with this data. Version 1.4 of the R library requires the data to
be in binary form. The require input for the R function is X representing the
individual level survey, dx representing the initial weights of the survey and
Tx representing the small area benchmarks.

```
# Warning: test code
# loop through simlaton areas
for(area in seq(dim(age)[1])){
    gregwt = GREGWT(data_in = data_in, area_code = area)
    fw <- gregwt$final_weights
    fweights <- c(fweights, fw)
    ## Estimate income
    sum.income <- sum(fw * ind$income)
    cap.income <- sum(fw * ind$income / sum(fw))
    Result[area,] <- c(area, sum.income, cap.income)
}
```

In the last step we transform the vector into a matrix and see the results from
the reweighing process.

6.2   Population synthesis as an optimization problem

In general terms, an *optimization problem* consists of a function, the result
of which must be minimised or maximised, called an *objective function*. This
function is not necessarily defined for the entire domain of possible inputs.
The domain where this function is defined is called the *solution space* (or the
feasible space in formal mathematics). Moreover, optimization problems can be
unconstrained or *constrained*, by limits on the values that arguments (or that
a function of the arguments) of the function can take (Boyd 2004 (http://
stanford.edu/~boyd/cvxbook/)). If there are constraints, the solution space
could include only a part of the image of the objective function. The objective
function and the constraints are both necessary to define the solution space.
Under this framework, population synthesis can be seen as a *constrained
optimisation* problem. Suppose x is a vector of length n $(x_1, x_2, .., x_n)$ whose

values are to be adjusted. In this case the value of the objective function is $f_0(x)$, depends on x. The possible values of x are defined thanks to *par*, a vector of predefined arguments or parameters of length m (m is the number of constraints) $(par_1, par_2, .., par_m)$. This kind of problem can be expressed as:

$$\begin{cases} min \ f_0(x_1, x_2, .., x_n) \\ s.c. \ f_i(x) \geq par_i, \ i = 1, ..., m \end{cases}$$

Applying this to the problem of population synthesis, the parameters par_i represent 0 and $f_i(x) = x$, since all cells have to be positive. The $f_0(x)$ to be minimised is the distance between the actual weight matrix and the aggregate constraint variable `cons`. x represents the weights which will be calculated to minimise $f_0(x)$.

To illustrate the concept further, consider the case of aircraft design. Imagine that the aim (the objective function) is to minimise weight by changing its shape and materials. But these modifications must proceed subject to some constraints, because the airplane must be safe and sufficiently voluminous to transport people. Constrained optimisation in this case would involve searching combinations of shape and material (to include in x) that minimise the weight (the result of f_0, is a single value depending on x). This search must take place under constraints relating to volume (depending on the shape) and safety (par_1 and par_2 in the above notation). Thus *par* values define the domain of the *solution space*. We search inside this domain for the combination of x_i that minimises weight.

The case of spatial microsimulation has relatively simple constraints: all weights must be positive or zero:

$$\{weight_{ij} \in \mathbb{R}^+ \cup \{0\} \quad \forall i, j\}$$

Seeing spatial microsimulation as an optimisation problem allows solutions to be found using established techniques of *constrained optimisation*. The main advantage of this reframing is that it allows any optimisation algorithm to perform the reweighting.

To see population synthesis as a constrained optimization problem analogous to aircraft design, we must define the problem to optimise the variable x and then set the constraints.

Intuitively, the weight (or number of occurrences) for each individual should be the one that best fits the constraints. We could take the weight matrix as x and as the objective function the difference between the population with this weight matrix and the constraint. However, we want to include the information of the distribution of the sample. We must find a vector `w` with which to multiply the `indu` matrix. `indu` is similar to `ind_cat`, but each row of `ind_cat` represents an individual of the sample, whereas each row of `indu` concerns a type of individual. This means that if 2 people in the sample have the same characteristics, the corresponding line in `indu` will appear only once. The cells of `indu` contain the number of times that this kind of individual appears in the sample. The result of this multiplication should be as close as possible to the constraints.

When running the IPF procedure zone-by-zone using this method, the optimization problem for the first zone can be written as follows:

$$\begin{cases} min \quad f(w_1, .., w_m) = DIST(sim, cons[1,]) \\ \qquad where \quad sim = colSums(indu * w) \\ \\ s.c. \quad w_i \geq 0, \quad i = 1, ..., m \end{cases}$$

Key to this is interpreting individual weights as parameters (the vector $w = (w_1, ..., w_m)$, of length m above) that are iteratively modified to optimise the fit between individual and aggregate level data. Note that in comparison with the theoretical definition of an optimisation problem, our parameters to determine (`par`) are the theoretical x. The measure of fit, so the distance, we use in this context is Total Absolute Error (TAE).

$$\begin{cases} min \quad f(w_1, .., w_m) = TAE(sim, cons[1,]) \\ \qquad where \quad sim = colSums(ind\_cat * w) \\ \\ s.c. \quad w_i \geq 0, \quad i = 1, ..., m \end{cases}$$

Note that although the "TAE" goodness-of-fit measure was used in this example, any could be used.

Note that in the above, w is equivalent to the `weights` object we have created in previous sections to represent how representative each individual is of each zone.

The main issue with this definition of reweighting is therefore the large number of free parameters: equal to the number of individual level dataset. Clearly this can be very very large. To overcome this issue, we must 'compress' the individual level dataset to its essence, to contain only unique individuals with respect to the constraint variables (*constraint-unique* individuals).

The challenge is to convert the binary 'model matrix' form of the individual level data (`ind_cat` in the previous examples) into a new matrix (`indu`) that has fewer rows of data. Information about the frequency of each constraint-unique individual is kept by increasing the value of the '1' entries for each column for the replicated individuals by the number of other individuals who share the same combination of attributes. This may sound quite simple, so let's use the example of SimpleWorld to illustrate the point.

6.2.1   Reweighting with optim and GenSA

The base R function `optim` provides a general purpose optimization framework for numerically solving objective functions. Based on the objective function for spatial microsimulation described above, we can use any general optimization algorithm for reweighting the individual level dataset. But which to use?

Different reweighting strategies are suitable in different contexts and there is no clear winner for every occasion. However, testing a range of strategy makes it clear that certain algorithms are more efficient than others for spatial microsimulation. Figure 6.1 demonstrates this variability by plotting total absolute error as a function of number of iterations for various optimization algorithms available from the base function `optim` and the **GenSA** package. Note that the comparisons are performed only for zone 1.

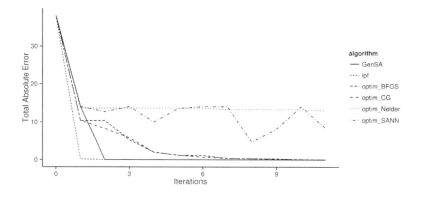

FIGURE 6.1
Relationship between number of iterations and goodness-of-fit between observed and simulated results for different optimisation algorithms.

Figure 6.1 shows that all algorithms improve fit during the first iteration. The reason for using IPF becomes clear after only one iteration. On the other end of the spectrum is R's default optimization algorithm, the Nelder-Mead method, which requires many more iterations to converge to a value approximating zero than does IPF.[1] Next best in terms of iterations is `GenSA`, the Generalized Simulated Annealing Function from the **GenSA** package. `GenSA` attained a near-perfect fit after only two full iterations.

---

[1] Although the graph shows no improvement from one iteration to the next for the Nelder-Mead algorithm, it should be stated that it is just 'warming up' at this stage and than each iteration is very fast, as we shall see. After 400 iterations (which happen in the same time that other algorithms take for a single iteration!), the Nelder-Mead begins to converge: it works effectively.

The remaining algorithms shown are, like Nelder-Mead, available from within R's default optimisation function `optim`. The implementations with `method =` set to `"BFGS"` (short for the Broyden–Fletcher–Goldfarb–Shanno algorithm), `"CG"` ('conjugate gradients') performed roughly the same, steadily approaching zero error and fitting to `"IPF"` and `"GenSA"` after 10 iterations. Finally, the `SANN` method (a variant of a Simulated ANNealing), also available in `optim`, performed most erratically of the methods tested. This is another implementation of simulated annealing which demonstrates that optimisation functions that depend on random numbers do not always lead to improved fit from one iteration to the next. If we look until 200 iterations, the fit will continue to oscillate and not be improved at all.

The code used to test these alternative methods for reweighting are provided in the script 'R/optim-tests-SimpleWorld.R'. The results should be reproducible on any computer, provided the book's supplementary materials have been downloaded. There are many other optimisation algorithms available in R through a wide range of packages and new and improved functions are being made available all the time. Enthusiastic readers are encouraged to experiment with the methods presented here: it is possible that an algorithm exists which outperforms all of those tested for this book. Also, it should be noted that the algorithms were tested on the extremely simple and rather contrived example dataset of SimpleWorld. Some algorithms may perform better with larger datasets than others and may be sensitive to changes to the initial conditions such as the problem of 'empty cells'.

Therefore these results, as with any modelling exercise, should be interpreted with a healthy dose of skepticism: just because an algorithm converges after few 'iterations' this does not mean it is inherently faster or more useful than another. The results are context specific, so it is recommended that the tested framework in 'R/optim-tests-SimpleWorld.R' is used as a basis for further tests on algorithm performance on the datasets you are using. IPF has performed well in the situations I have tested it in (especially via the `ipfp` function, which performs disproportionately faster than the pure R implementation on large datasets) but this does not mean that it is always the best approach.

To overcome the caveat that the meaning of an 'iteration' changes dramatically from one algorithm to the next, further tests measured the time taken for each reweighting algorithm to run. To have a readable graph, we do not represent the error as a function of the time, but the time per algorithm in function of the number of iterations (Figure 6.2). This figure demonstrates that an iteration of GenSA take a long time in comparison with the other algorithm. Moreover, `"BFGS"` and `"CG"` are still following a similar curve under GenSA. Nelder-Mead, SANN and IPF contains iterations that take less time. By observing Figures 6.1 and 6.2 simultaneously , it appears that IPF is the best in terms of convergence (little TAE after few iterations) and the time needed for few iterations is good.

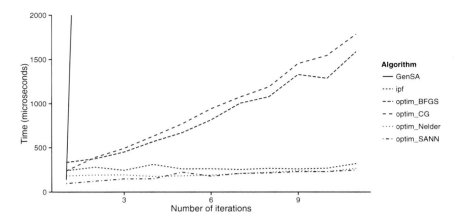

FIGURE 6.2
Relationship between processing time and goodness-of-fit between observed and simulated results for different optimisation algorithms.

Nelder-Mead is fast at reaching a good approximation of the constraint data, despite taking many iterations. `GenSA`, on the other hand, is shown to be much slower than the others, despite only requiring 2 iterations to arrive at a good level of fit.

Note that these results are biased by the example that is pretty small and runs only for the first zone.

6.2.2   Combinatorial optimisation

Combinatorial optimisation (CO) is a set of methods for solving optimisation problems from a discrete range of options. Instead of allocating weights to individuals per zone, combinatorial optimisation identifies a set of feasible candidates to 'fill' the zone and then swaps in new individuals. The objective function to be minimised is used to decide if the swap was beneficial or not. There are many combinatorial optimisation methods available. Which is suitable depends on how we choose the combination of candidates and how we determine what happened after evaluation.

CO is an alternative to IPF for allocating individuals to zones. This strategy is probabilistic and results in integer weights (since it is a combination of individuals). Combinatorial optimisation may be more appropriate for applications where input individual microdatasets are very large: the speed benefits of using the deterministic IPF algorithm shrink as the size of the survey dataset increases. As seen before, IPF creates non integer weights, but we have proposed two solutions to transform them into the final individual level population. So, the proportionality of IPF is more intuitive, but need to calculate the whole weight matrix at each iteration, where CO just proposes candidates. If the objective function takes a long time to be calculated CO can be computationally intensive because the goodness-of-fit must be evaluated each time a new population is proposed with one or several swapped individuals.

Genetic algorithms are included in this field and become popular in some domains, such as industry, for the moment. This kind of algorithm can be very effective when the objective function has several local minima and we want to find the global one (Hermes and Poulsen, 2012).

To illustrate how the integer-based approach works in general terms, we can use the `data.type.int` argument of the `genoud` function in the **rgenoud** package. This ensures only integers result from the optimisation process:

```
# Set min and maximum values of constraints with 'Domains'
m <- matrix(c(0, 100), ncol = 2)[rep(1, nrow(ind)),]
set.seed(2014)
genoud(nrow(ind), fn = fun, ind_num = ind, con = cons[1,],
    control = list(maxit = 1000), data.type.int = TRUE, D = m)
```

This command, implemented in the file 'R/optim-tests-SimpleWorld.R', results in weights for the unique individuals 1 to 4 of 1, 4, 2 and 4 respectively. This means a final population with aggregated data equal to the target (as seen in the previous section):

```
## ind$agea0_49   ind$agea50+    ind$sexm      ind$sexf
##            8             4           6             6
```

Note that we performed the test only for zone 1 and the aggregated results are correct for the first constraint. Moreover, thanks to the fact that the algorithm only considers integer weights, we do not have the issue of fractional weights associated with IPF. Combinatorial optimisation algorithms for population synthesis do not rely on integerisation, which can damage model fit. The fact that the gradient contains "NA"[2] in the end of the algorithm is not a problem, because it just means that this cell has not been calculated.

Note that there can be several solutions which attain a perfect fit. This result depends on the random seed chosen for the random draw. Indeed, if we chose a seed of 0 (by writing `set.seed(0)`), as before, we obtain the weights (0, 6, 4, 2) which results also in a perfect fit for zone 1. These two potential synthetic populations reach a perfect fit, but are quite different. Indeed, we can observe the two populations.

---

[2]We mean that instead of having a numerical value in each cells, some cells contain the value 'NA'(Non Applicable). This value appears when this cell has not been defined.

An example of comparison is that the second proposition contains no male being more than 50 years old, but the first one has 2. With this method, there cannot be a population with 1 male of over 50, because we take integer weights and there are two men in this category in the sample. This is the disadvantage of algorithms reaching directly integer weights. With IPF, if the weights of this individual are between 0 and 1, there is a possibility of a person belonging to this category.

genoud is used here only to provide a practical demonstration of the possibilities of combinatorial optimisation using existing R packages.

For combinatorial optimisation algorithms designed for spatial microsimulation we must, for now, look for programs outside the R 'ecosystem'. Harland (2013) provides a practical tutorial introducing the subject using the Java-based Flexible Modelling Framework (FMF).

6.3   simPop

The **simPop** package provides alternative methods for generating and modelling synthetic microdata. A useful feature of the package is its inclusion of example data. Datasets from the 'EU-SILC' database (EU statistics on income and living conditions) and the developing world are included to demonstrate the methods.

An example of the individual level data provided by the package, and the function contingencyWt, is demonstrated below.

```
# install.packages("simPop")
library(simPop)

# Load individual level data
data("eusilcS")
head(eusilcS)[2:8]
```

```
##        hsize        db040 age   rb090 pl030 pb220a netIncome
## 9292       2     Salzburg  72    male     5     AT  22675.48
## 9293       2     Salzburg  66  female     5     AT  16999.29
## 7227       1 Upper Austria  56  female     2     AT  19274.21
## 5275       1       Styria  67  female     5     AT  13319.13
## 7866       3 Upper Austria  70  female     5     AT  14365.57
## 7867       3 Upper Austria  46    male     3     AT      0.00
```

```
# Compute contingency coefficient between two variables
contingencyWt(eusilcS$pl030, eusilcS$pb220a,
  weights = eusilcS$rb050)
```

```
## [1] 0.1749834
```

The above code shows the first 8 (of 18) variables in the `eusilcS` dataset provided by **simPop** and an example of one of **simPop**'s functions. This function is useful, as it calculates the level of association between two categorical variables: economic status (`pl030`) and citizenship status (`pb220a`). The result of 0.175 shows that there is a weak association between the two variables. **simPop** is a new package so we have not covered it in detail. However, it is worth considering for its provision of example datasets alone.

Note that this package provides several alternative methods for population synthesis, including "model-based methods, calibration and combinatorial optimization algorithms" (Meindl et al., 2015).

6.4 The Urban Data Science Toolkit (UDST)

The UDST provides an even more ambitious approach to modelling cities that contains a population synthesis component. An online overview of the project - http://www.udst.org/ - shows that the developers of the UDST are interested in visualisation and modelling, not just population synthesis. The project is an evolving open source project hosted on GitHub. It is outside the scope of this book to comment on the performance of the tools within the UDST (which is written in Python). Suffice to flag the project and suggest readers investigate the Synthpop (https://github.com/UDST/synthpop) software as an alternative synthetic population generator.

6.5 Chapter summary

In summary, this chapter presented several alternatives to IPF to generate spatial microdata. We explored a regression method (GREGWT), combinatorial optimization and the R package *simPop*. Finally we looked at the Python-based Urban Data Science Toolkit.

7

Spatial microsimulation in the wild

CONTENTS

So far the book has explained what spatial microsimulation is, described its applications and demonstrated how it works. We have seen something of its underlying theory and its implementation in R. But how can the method be applied 'in the wild', on real datasets?

The purpose of this chapter is to answer this question using real data to estimate cake consumption in different parts of Leeds, UK. The example is deliberately rather absurd to make it more memorable. The steps are presented in a generalisable way, to be applicable to a wide range of datasets.

The input microdataset is a randomized ('jumbled') subset of the 2009 Dental Health Survey (http://data.gov.uk/dataset/adult_dental_health_survey), (DHS) which covers England, Northern Ireland and Wales. 1173 variables are available in the DHS, many of which are potentially interesting target variables not available at the local level. These include weekly income, plaque build-up, and oral health behaviours. Potential linking variables include socio-economic classification, and dozens of variables related to oral health.

In terms of constraint variables, we are more limited: the Census is the only survey that provides count data at the small area level in the UK. Thus the 'domain' of available input data, related to our research question involves two sources:

1. Non-geographical individual level survey, DHS — the *microdata.*
2. Geographically aggregated categorical count data from the census — the *constraint tables.*

This chapter is structured as follow:

- *Selection of constraint variables* (Section 7.1) describes and justifies the choice of the constraint variables on this specific example.
- *Preparing the input data* (Section 7.2) contains the pieces of code necessary to load and prepare the data, before performing the spatial microsimulation.
- *Using the ipfp package* (Section 7.3) develops the whole process to create a synthetic population thanks to the R package **ipfp**.
- *Using the mipfp package* (Section 7.4) also includes the whole process, but with the R package **mipfp**.
- *Comparing methods of reweighting large datasets* (Section 7.5) compares the different results and describes briefly how to transform one type of result to the other type.

7.1   Selection of constraint variables

As discussed in Chapter 4, we must first decide which variables should be used to link the two. We must select the constraints from available linking variables.

The selection of linking variables should not be arbitrarily preordained by preconceptions. The decision of which constraints to use to allocate individuals to zones should be context dependent. If the research is on social exclusion, for example, many variables could potentially be of interest: car ownership, house tenancy, age, gender and religion could all affect the dependent variable. Often constraint variables must be decided not based on what would be ideal, but which datasets are available. The selection criteria will vary from one project to the next, but there are some overriding principles that apply to most projects:

1. **More the merrier**: each additional constraint used will further differentiate the spatial microdata from the input microdata. If gender is the only constraint used, for example, the spatial microdata will simply be a repetition of the input microdata but with small differences in the gender ratio from one zone to the next. If five constraints are used (e.g. age, gender, car ownership, tenancy and religion), the differences between the spatial microdata from one zone to the next will be much more pronounced and probably useful.

2. **Relevance to the target variable**: often spatial microsimulation is used to generate local estimates of variables about which little geographically disaggregated information is available. Income is a common example: we have much information about income distributions, but little information about how average values (let alone the distribution) of income varies from one small area to the next. In this case income is the target variable. Therefore constraints must be selected which are closely related to income for the output to resemble reality. This is analogous to multiple regression (which can also be used to estimate average income at the local level), where the correct *explanatory variables* (i.e. constraint variables in spatial microsimulation) must be selected to effectively predict the *dependent variable*. As with regression models, there are techniques which can be used to identify the most suitable constraint variables for a given target variable.

3. **Simplicity**: this criterion to some extent contradicts the first. Sometimes more constraints do not result in better spatial microdata and problems associated with 'over-fitting' can emerge. Spatial microsimulation models based on many tens of constraint categories will take longer to run and require more time to develop and modify. In addition, the chances of an error being introduced during every phase of the project is increased with each additional constraint. The extent to which increasing the number of constraint categories improves the results of spatial microsimulation, either with additional variables or by using cross-tabulated constraints (e.g. age/sex) instead of single-variable constraints, has yet to be explored. It is therefore difficult to provide general rules of thumb regarding simplicity other than 'do not over-complicate the model with excessive constraint variables and constraints'.

So, we always need to reach an equilibrium between these principles. Indeed, we have to take into account *enough* and *pertinent* variables, without making the population synthesis process too complex (see Section 4.2).

To exemplify these principles, let us consider the constraint variables available in the CakeMap datasets. Clearly only variables available both in the individual level and aggregate level datasets can be chosen. Suppose our aim is to analyse the consumption of cakes depending on socio-demographic variables. Five interesting variables assigned to each of the 916 individuals are available from the individual level data:

- 'Car': The number of working cars in the person's household.
- 'Sex' and 'ageband4': Gender and age group, in two separate variables. Age is divided into 6 groups ranging from '16–24' to '65–74.'[1]
- 'NSSEC': National Statistics Socio-economic Classification: a categorical variable classifying the individual's work into one of 10 groups including '97', which means 'no answer' (NA).
- 'NCakes': the target variable, reported number of times that the respondent consumes cake each week.

All of these variables, except for 'NCakes', have a corresponding constraint variable to be loaded for the 124 Wards that constitute the Leeds Local Authority in the UK. In real datasets it is rarely the case that the categories of the individual and aggregate level data match perfectly from the outset. This is the first problem we must overcome before running a spatial microsimulation model of cake consumption in Leeds.

The code needed to run the main part of the example is contained within 'CakeMap.R'. Note that this script makes frequent reference to files contained in the folder 'data/CakeMap', where input data and processing scripts for the project are stored.

7.2   Preparing the input data

Often spatial microsimulation is presented in a way that suggests the data arrived in a near perfect state, ready to be inserted directly into the model. This is rarely the case; usually, one must spend time loading the data into R, dealing with missing values, re-coding categorical variables and column names, binning continuous variables and subsetting from the microdataset. In a typical project, data preparation can take as long as the analysis stage. This section builds on Chapter 2 to illustrate strategies for data cleaning on a complex project. To learn about the data cleaning steps that may be useful to your data, we start from the beginning in this section, with a real (anonymised) dataset that was downloaded from the internet.

The raw constraint variables for CakeMap were downloaded from the Infuse website (http://infuse.mimas.ac.uk/). These, logically enough, are stored in the 'cakeMap/data/' directory as .csv files and contain the word 'raw' in the file name to identify the original data. The file 'age-sex-raw.csv', for example,

---

[1]R tip: This information can be seen, once the dataset is loaded, by entering unique(ind$ageband4) or, to see the counts in each category, summary(ind$ageband4). Because the variable is of type factor, levels(ind$ageband4) will also provide this information.

is the raw age and sex data that was downloaded. As the screenshot in Figure 7.1 illustrates, these datasets are rather verbose and require pre-processing. The resulting 'clean' constraints are saved in files such as 'con1.csv', which stands for 'constraint 1'.

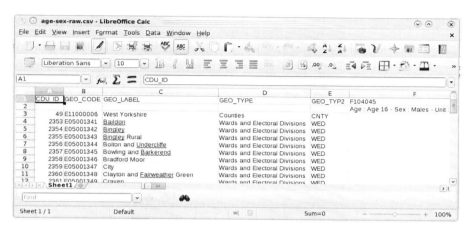

FIGURE 7.1
Example of raw aggregate level input data for CakeMap aggregate data, downloaded from `http://infuse.mimas.ac.uk/`.

To ensure reproducibility in the process of converting the raw data into a form ready for spatial microsimulation, all the steps have been saved. Take a look at the R script files 'process-age.R', 'process-nssec.R' and 'process-car.R'. The contents of these scripts should provide insight into methods for data preparation in R. Wickham (2014b) provides a more general introduction to data reformatting. The most difficult input dataset to deal with, in this example, is the age/sex constraint data. The steps used to clean it are saved in 'process-age.R', in the **data/CakeMap/** folder. Take a look through this file and try to work out what is going on: the critical stage is grouping single year age bands into larger groups such as 16–24.

The end result of 'process-age.R' is a 'clean' .csv file, ready to be loaded and used as the input of our spatial microsimulation model. Note that the last line of 'process-age.R' is `write.csv(con1, "con1.csv", row.names = F)`. This is the first constraint that we load into R to reweight the individual level data in the next section. The outputs from these data preparation steps are named 'con1.csv' to 'con3.csv'. For simplicity, all these were merged (by 'load-all.R') into a single dataset called 'cons.csv'. All the input data for this section are thus loaded with only two lines of code:

```
ind <- read.csv("data/CakeMap/ind.csv")
cons <- read.csv("data/CakeMap/cons.csv")
```

Take a look at these input data using the techniques learned in the previous section. To test your understanding, try to answer the following questions:

- What are the constraint variables?
- How many individuals are in the survey microdataset?
- How many zones will we generate spatial microdata for?

For bonus points that will test your R skills as well as your practical knowledge of spatial microsimulation, try constructing queries in R that will automatically answer these questions.

It is vital to understand the input datasets before trying to model them, so take some time exploring the input. Only when these datasets make sense (a pen and paper can help here, as well as R!) is it time to generate the spatial microdata.

As explained in the previous chapters, there are different methods to perform a spatial microsimulation. We will here use a reweighting method, since we have spatial aggregated data and non spatial individual data, both containing some common variables. In this category, the more often used and intuitive method is the IPF.

We mentioned in the SimpleWorld example that there are two different points of view of IPF. First, weights are assigned to each individual for each zone (`ipfp` package). Second, weights are assigned to each possible category of individuals for each zone (`mipfp` package). The next sections develop the whole procedure with each package.

7.3   Using the `ipfp` package

7.3.1   Performing IPF on CakeMap data

The `ipfp` reweighting strategy is concise, generalisable and computationally efficient. On a modern laptop, the `ipfp` method was found to be *almost 40 times faster* than the 'IPFinR' method (section 4.1; Lovelace, 2014) over 20 iterations on the CakeMap data, completing in 2 seconds instead of over 1 minute. This is a huge time saving![2]

---

[2]These tests were conducted using the `microbenchmark()` commands found towards the end of the 'CakeMap.R' file. The second of these benchmarks depends on files from `smsim-course` (Lovelace, 2014), the repository of which can be downloaded from (`https://github.com/Robinlovelace/smsim-course`).

Thanks to the preparatory steps described above, the IPF stage can be run on a single line. After the datasets are loaded in the first half of 'CakeMap.R', the following code creates the weight matrix:

```
weights <- apply(cons, 1, function(x)
  ipfp(x, ind_catt, x0, maxit = 20))
```

As with the SimpleWorld example, the correlation[3] between the constraint table and the aggregated results of the spatial microsimulation can be checked to ensure that the reweighting process has worked correctly. This demonstrates that the process has worked with an r value above 0.99:

```
cor(as.numeric(cons), as.numeric(ind_agg))
```

```
## [1] 0.9968529
```

This value is very close to 1, so we can consider that there is a big linear correlation between the results and the constraints. We can verify it by plotting the graph on Figure 7.2.

With the best fit, we would have all constraints equal to the simulation, so a perfect line (this is why we consider a `linear` correlation). Only few points are outside the area of the line. Note that we have 1,623,797 inhabitants[4] in the simulation and in the constraints, `con1` and `con2` have 1,623,800, while `con3` contains 1,623,797. As seen before, with IPF, the result depends on the order of the constraints. Thus, it is logical to put as last constraint one really reliable, also in terms of total number of people. An alternative is to rescale all constraints to be consistent. This will be done in the next section, with `mipfp` and in the comparison for `ipfp`.

For the comparison category per category, we can take the absolute value of the differences of the two tables. This gives a table of differences. The worst category and zone will be the maximum of this matrix.

```
# Maximum error
max(abs(ind_agg-cons))
```

```
## [1] 4960.299
```

---

[3]The function "cor" of R calculates the correlation coefficient of Pearson that measures the force of a linear correlation between the two variables. Included between -1 and 1, the best value when comparing a simulation of a theoretical count (this is the case here) is 1.

[4]To obtain it, code `sum(ind_agg)/3`, because you have 3 constraints.

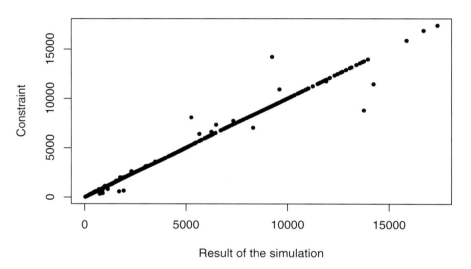

FIGURE 7.2
Scatter plot of the relationship between observed cell counts for all categories
and all zones from the census (x axis) and the simulated cell counts after IPF
(y axis).

```
# Index of the maximum error
which(abs(ind_agg-cons) == max(abs(ind_agg-cons)),
      arr.ind = TRUE)
```

```
##      row col
## [1,]  84  13
```

The maximum error is now known and corresponds to the zone 84 and 13 columns of the constraints, which is `Car`.

This is for the basic validation; a more detailed analysis of the quality of the results is present in the next chapter.

7.3.2 Integerisation

As before, weights of the IPF procedure are fractional, so must be *integerised* to create whole individuals. The code presented in Chapter 4 requires little modification to do this: it is your task to convert the weight matrix generated by the above lines of code into a spatial microdataset called, as before, `ints_df` (hint: the `int_trs` function in 'R/functions.R' file will help). The spatial microdata generated in 'R/CakeMapInts.R' contains the same information as the individual level dataset, but with the addition of the 'zone' variable, which specifies which zone each individual inhabits.

The spatial microdata is thus *multilevel* data, operating at one level on the level of individuals and at another on the level of zones. To generate summary statistics about the individuals in each zone, functions must be run on the data, one zone (group) at a time. `aggregate` provides one way of doing this. After converting the 'NSSEC' socio-economic class variable into a suitable numeric variable, `aggregate` can be used to identify the variability in social class in each zone. The `by =` argument is used to specify how the results are grouped depending on the zone each individual inhabits:

```
source("R/CakeMapInts.R")
```

```
aggregate(ints_df$NSSEC, by = list(ints_df$zone), sd,
    na.rm = TRUE)
```

```
##   Group.1        x
## 1       1 1.970570
## 2       2 2.027638
## 3       3 2.019839
```

In the above code the third argument refers to the function to be applied to the input data. The fourth argument is simply an argument of this function, in this case instructing the standard deviation function (`sd`) to ignore all `NA` values. An alternative way to perform this operation, which is faster and more concise, is using `tapply`:

```
tapply(ints_df$NSSEC, ints_df$zone, sd, na.rm = TRUE)
```

Note that operations on `ints_df` can take a few seconds to complete. This is because the object is large, taking up much RAM on the computer. This can be seen by asking `object.size(ints_df)` or `nrow(ints_df)`. The latter shows we have created a spatial microdataset of 1.6 million individuals! Try comparing this result with the size of the original survey dataset 'ind'. Keeping an eye on such parameters will ensure that the model does not generate datasets too large to handle.

Next chapter, we move on to a vital consideration in spatial microsimulation models such as CakeMap: model checking and validation. First, we develop the spatial microsimulation using `mipfp`.

7.4   Using the `mipfp` package

7.4.1   Performing IPF on CakeMap data

As in the previous section, the first stage is to load the data. An additional step required by the `Ipfp` function is ensuring the categories of `cons` and `ind` correspond. In the variable associated with car ownership, `ind` contains "1" and "2", but `cons` contains "Car" and "NoCar".

To undertake this operation, we convert the category labels in the individual level variable `ind` so that they conform to the variable names of the count data contained in constraint data `cons`. Continuing with the car ownership example, every "1" in `ind$Car` should be converted to "Car" and every "2" to "NoCar". Likewise, the correct labels of "f" and "m" must be created for the Sex variable and likewise for the other variables in `ind`. The code that performs this operation is included in the R file `CakeMapMipfpData.R` file. Take a look at this file (e.g. by typing `file.edit("R/CakeMapMipfpData.R")` from RStudio) and note the use of the `switch` function in combination with the recursive function `sapply` to *switch* the values of the individual level variables. The entire script is run below using the `source` command.

```
source("R/CakeMapMipfpData.R")
```

The data are now loaded correctly. This can be verified by printing the first lines of the individual level data and comparing this with colnames(con1). Note that this step was not necessary for the **ipfp** package. To use **ipfp** the user must encode the constraints in the right order and only the order counts. In practice this is a good idea. Ensuring same variable names match category labels will help subsequent steps of validation and interpretation. Moreover, the structure of some constraints will have to change (**mipfp** takes as input a table of n_var dimensions, whereas **ipfp** takes as input a 2-dimensional dimension, whose columns describe variables) and this process will be easier if the names are the same. To reaffirm our starting point, a sample of the new individual level data is displayed below.

```
head(ind)
```

```
##    NCakes   Car Sex NSSEC8 ageband4
## 1    3-5 NoCar   m     X7    25_34
## 2    1-2 NoCar   f     X2    55_64
## 3    1-2   Car   f     X2    45_54
## 4     6+   Car   m     X5    45_54
## 5    1-2 NoCar   m     X2    45_54
## 6    1-2   Car   f     X2    45_54
```

Defining the initial weight matrix is the next step. For each zone, we simply take the cross table (or contingency table) of the individual data. It is important to be aware of the order of the variables inside this table. We have in our case (NCakes, Car, Sex, NSSEC8, ageband4). This information will be necessary for the definition of the target and for the description for mipfp.

```
# Initial weight matrix for each zone
weight_init_1zone <- table(ind)

# Check order of the variables
dimnames(weight_init_1zone)
```

```
## $NCakes
## [1] "<1"      "1-2"     "3-5"     "6+"      "rarely"
##
## $Car
## [1] "Car"     "NoCar"
##
## $Sex
```

```
## [1] "f" "m"
##
## $NSSEC8
##  [1] "X1.1"  "X1.2"  "X2"   "X3"   "X4"   "X5"   "X6"   "X7"
##  [9] "X8"    "Other"
##
## $ageband4
## [1] "16_24" "25_34" "35_44" "45_54" "55_64" "65_74"
```

To use the full power of the `mipfp` package, we choose to add the spatial
dimension into the weight matrix instead of performing a `for` loop over the
zones. For this purpose, we repeat the table for each area. We define the names
of the new dimension and create an array with the correct cells and names.

```
# Adding the spatial dimension
# Repeat the initial matrix n_zone times
init_cells <- rep(weight_init_1zone, each = nrow(cons))

# Define the names
names <- c(list(rownames(cons)),
           as.list(dimnames(weight_init_1zone)))

# Structure the data
weight_init <- array(init_cells, dim =
                  c(nrow(cons), dim(weight_init_1zone)),
                  dimnames = names)
```

An advantage of **mipfp** over **ipfp** is that the algorithm first checks the consistency of the constraints before reweighting. If the total number of individuals in
a zone is not exactly the same from one constraint to another, `mipfp` will return
the following warning: 'Target not consistent - shifting to probabilities!'. In this
case, the cells of the resulting table become probabilities, such as the sum of
all cells is 1. This means that the algorithm does not know how many people
per zone you want. This is a very useful feature of **mipfp**: creating weights
corresponding to the correct total population is as simple as multiplying the
probabilities by this number.

An alternative to this is to verify by your own the totals per zone and re-scale
the constraints that are different from your purpose population. In our case,
for each constraint, we calculate the number of persons in the zone. For each
zone, we compare the totals of the different constraints. The result is `TRUE`
if the totals are the same and `FALSE` otherwise. It gives `nrow(cons)` boolean
values (TRUE or FALSE), of which we print the table.

```
# Check the constraint's totals per zone
table(rowSums(con2) == rowSums(con1))
```

```
##
## TRUE
##  124
```

```
table(rowSums(con3) == rowSums(con1))
```

```
##
## FALSE   TRUE
##    72     52
```

```
table(rowSums(con2) == rowSums(con3))
```

```
##
## FALSE   TRUE
##    72     52
```

Constraints 1 and 2 have exactly the same marginals per zone. Constraint 3 has 72 zones with different totals. Since NSSEC (con3) is anomalous in this respect we consider the totals of the two first constraints to be valid. It is always important to find the reason of the differences in totals. For con3, the individual could give no answer. This explains why we have a smaller population in this variable. To perform (non-probabilistic) reweighting with **mipfp**, we must first rescale con3 to keep its proportions but with updated totals per zone.

```
# Re-scale the constraint 3
con3_prop <- con3 * rowSums(con2) / rowSums(con3)
```

```
# Check the new totals
table(rowSums(con2) == rowSums(con3_prop))
```

```
##
## TRUE
##  124
```

The rescaling operation above solved one problem, but there is another, more difficult-to-solve issue with the input data. The second and third constraints are simple marginals per zone with a single, clearly-defined variable each. The first constraint (con1), by contrast, is cross-tabulated, a combination of two

variables (sex and age[5]). We must change the structure of the first constraint
to execute `mipfp`.

```
# View the content of the first constraint
con1[1:3, 1:8]
```

```
##    m16_24 m25_34 m35_44 m45_54 m55_64 m65_74 f16_24 f25_34
## 1     671    771   1033   1160   1165    772    679    760
## 2     887   1254   1217   1344   1229    752    832   1169
## 3     883    864   1195   1382   1170    878    811    962
```

The results of the above command show that `con1` is, like the other constraints,
a two–dimensional table. Because it actually contains 2 constraints per zone, the
information contained within needs to be represented as a three–dimensional
array (a cube), the dimensions of which are zone, sex and age. This conversion is
done by taking the cells of `con1` and putting them to a three–dimensional array
in the correct order (corresponding to the order of the 2–dimensional table).
The code to do this is included into the file `R/CakeMapMipfpCon1Convert.R`
and the resulting array is called `con1_convert`.

```
source("R/CakeMapMipfpCon1Convert.R")
```

All constraints being converted and all the margins being rescaled, the target
and its 'description' are ready to be used in `Ipfp()`. The target is the set of
constraints and the description is the figures of the corresponding dimensions
into the weight matrix. The order of the weight's variables is here important.
Remember this order is (Zone, NCakes, Car, Sex, NSSEC8, ageband4). For this
reason, the constraint 1, containing (zone, Sex, ageband4) has the description
`c(1,4,6)`.

```
# Definition of the target and descript
# For the execution of the mipfp
target <- list(con1_convert,
               as.matrix(con2),
               as.matrix(con3_prop))

descript <- list(c(1,4,6), c(1,3), c(1,5))
```

The remaining step involves executing the `Ipfp` function on these data.

---

[5]or age × sex, as cross-tabulations are often represented.

```
weight_mipfp <- Ipfp(weight_init, descript, target)
```

Note that `Ipfp` does not print the `tol` result by default. This was a wise decision by the developer of **mipfp** in this case: more than 100 iterations were needed for the weights to converge.

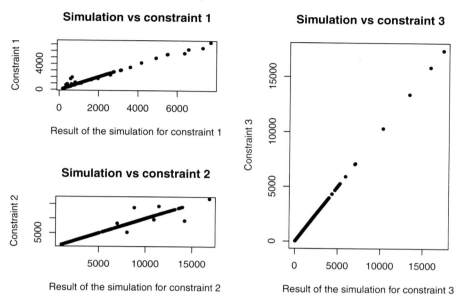

FIGURE 7.3
Scatter plot of the relationship between observed cell counts for all categories and all zones from the census (x axis) and the simulated cell counts after IPF (y axis).

We can see on Figure 7.3 that the third constraint is perfectly fitted. The first and second constraints are globally well fitted, but some points, representing a specific category in the specific zone, are not so good. Analysis of quality of fit is provided in the next chapter. In the next section we focus on comparing **ipfp** and **mipfp** for real-world microsimulation applications involving large datasets, such as CakeMap.

7.5 Comparing methods of reweighting large datasets

We mentioned in the **ipfp** part of this chapter that the three constraints do not contain the same number of individuals. To compare both methods, we will run a second time `ipfp`, but with the rescaled constraints. Note that in case of non-consistent constraints, `mipfp` warns the user, but `ipfp` does not. Thus, with `ipfp`, it is recommended that the user checks constraint totals before reweighting. Note that instead of the number of iterations, we chose to fix the tolerance threshold to 10^{-10}, the default tolerance argument used in the `mipfp` function (remember, such details can be verified by typing `?mipfp`).

```
# Re-running ipfp with the rescaled constraints
cons_prop <- cbind(con1, con2, con3_prop)
weights <- apply(cons_prop, 1,
                 function(x) ipfp(x, ind_catt, x0, tol = 1e-10))

# Convert back to aggregates
ind_agg <- t(apply(weights, 2, function(x) colSums(x * ind_cat)))
```

This section is divided in two parts, one that analyses the distance between the results generated by the two packages and the other that focus on the time differences.

7.5.1   Comparison of results

Both results being based on the rescaled data, we can compare them. The structure of both resulting tables are different, however. To compare them easily, we transform the `ipfp` final matrix into an array with the same structure as the one of the result of `mipfp`.

```
source("R/ConvertIpfpWeights.R")
```

The total number of people per zone is exactly the same in both weights matrices (maximum of differences being of order 10^{-10}). The size of the population is in both cases the one desired[6].

```
# Comparison of the totals per zone
max( apply(weight_mipfp$x.hat, 1, sum)
       - apply(weight_ipfp, 1, sum) )
```

```
## [1] 7.275958e-12
```

To visualize all the weights, we plot the different counts per constraint at Figure 7.4. We observe that the constraint 3 is very well fitted with both methods. For the two other constraints, some categories in some zones are

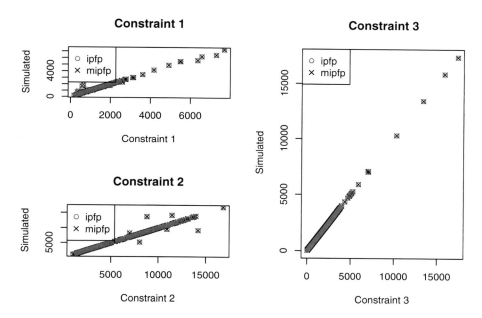

FIGURE 7.4
Scatter plot of the relationship between observed cell counts for all categories and all zones from the census (x axis) and the simulated cell counts after IPF (y axis) (ipfp vs mipfp).

more apart. However, both methods seems to return very similar weights, since the crosses are included inside the circles.

By analysing the absolute differences between the two resulting weights, we discover that the largest absolute distance is 3.45608e-11.

We can consider that both methods give very similar results, even if there are both distant from the constraints for few zones and categories. A longer analysis of this is done in the next chapter.

7.5.2 Comparison of times

The weights produced by **ipfp** and **mipfp** packages are the same. Thus, we must use other criteria to assess which is more appropriate in different settings. For applications involving very large microdatasets and numbers of zones, the speed of execution may be an important criterion. For this reason, the time is an important characteristic to compare the packages. However, other considerations such as flexibility and user-friendliness should also be taken into account. The question of 'scalability' (how large can the simulation be before the software crashes) is also important.

When reweighting individuals for the case of CakeMap example, `mipfp` was slower than `ipfp`. However, when accounting for wider considerations (for example if we replicate the zones two times) the speed advantage of `ipfp` is reduced. Tests show that as the size of input microdata increases, any performance advantage of `ipfp` declines. This phenomenon should not appear in the `mipfp` package. To demonstrate this we will run the CakeMap example with variable numbers of individuals in the microdata. To do this, **ind** objects with more than 916 persons in the data are created, by copying the database a second time. The code used to create new **ind** objects can be seen by entering `file.edit("R/CakeMapTimeAnalysis.R")` in RStudio from inside the 'spatial-microsimulation-book' project.

Figure 7.5 confirms that for problems with few individuals in the microdata, `ipfp` is better. However, having few people means the increase of the probability to have non–represented categories. Of course, we can optimise `ipfp` to take only one time the individuals that appear several times and add a count in the `ind_cat` instead of a "1" (as made in the **indu** variable of SimpleWorld). However, this process is possible only if the added persons are with the same characteristics as a previous one.

In conclusion, `mipfp` is more flexible and could be used without microdata, but `ipfp` is faster for problems with less individuals in the microdata.

---

[6]Note that to affirm this, we checked the total population in the constraints.

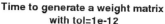

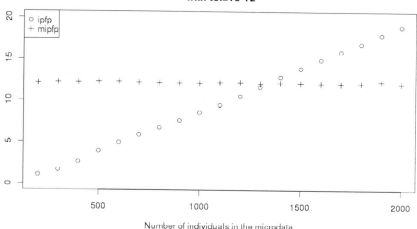

FIGURE 7.5
Time necessary to perform the generation of the weight matrix depending on the number of individuals inside the microdata.

These results suggest that to perform population synthesis for spatial microsimulation in contexts where the input microdataset is large, one should use `mipfp`. Note that when having very few microdata, it may be advantageous to consider a method without microdata. This way avoids biasing the results with the (non-random) categories included in the microdata.

The extent to which the input microdata represent the study area is critical to obtaining valid results: the results of spatial microsimulation that use input microdata are highly dependent on the extent to which the microdata are representative of the study area (Lenormand and Deffuant, 2012).

If the used sample is too small, or if the sample, for example, contains 90% female, whereas the whole population has only 51%, the microdata will not be representative. Usually, to consider a sample as representative, each individual should have exactly the same probability of being in the sample. For example, if you make your survey in a university, older people have less chance to be inside than younger. In this case, the sample is not representative.

7.6   Chapter summary

In summary, this chapter has coded the whole population synthesis for the CakeMap example. This reasoning has been developed with two R packages: **ipfp** and **mipfp**. This allowed us to observe the differences in the structure of the input data:

1. **ipfp** takes as input the individual data of the sample and the counts of the constraints included in a two–dimensional array (even if we have several constraint variables)
2. **mipfp** considers input in contingency tables. This means that if you have 5 different table of constraints, each included cross–tabulated variables, **mipfp** will take into account the 5 tables.

As illustrated in the chapter, we can transform the input of one algorithm into the input of the other, but **mipfp** is really simpler to use when having several and cross-tabulated constraints.

We have also seen that **ipfp** is faster for very small problems, but **mipfp** performs better on large datasets.

8

Model checking and evaluation

CONTENTS

In food safety, openness about mistakes is a vital ingredient for high standards.[1] The same concept applies to modelling. Transparency in model evaluation — the process of deciding whether the model is appropriate and identifying *how good* the results are — is vital in spatial microsimulation for similar reasons. Openness of code and method, as demonstrated and advocated throughout this book, is easy using command-line open source software such as R (Wickham 2014) (`http://adv-r.had.co.nz/Reproducibility.html`).

Reproducibility is especially important during model checking and evaluation, allowing you and others not only to *believe* that the model is working, but to *affirm* that the results are as expected. This chapter is about specific methods to check and evaluate the outputs of spatial microsimulation. The aims are simple: to ensure that the models 1) make sense given the input data (*model checking*) and 2) coincide with external reality (*model evaluation*). These strategies are described below:

1. *Model checking* — also called *internal validation* (Edwards et al. 2010) — is comparing model results against a priori knowledge of how they *should* be. This level of model checking usually takes place only at the aggregate level of the constraint variables (Section 8.1).

---

[1] This seems to be because hiding or being ashamed of inevitable mistakes allows bad practice to continue unnoticed (Powell et al. 2011).

2. *Model evaluation* — also known as *external validation* — is the process of comparing model results with external data. This approach to verification relies on good 'ground-truth data' and can take place at either the individual level (if geo-coded survey data are available) or (more commonly) at the aggregate level (Section 8.3).

The chapter develops these two possible validations and explains the problem of the zero cells (Section 8.2).

Internal validation is the most common form of model evaluation. In some cases this type of validation is the only available test of the model's output, because datasets for external validation are unavailable. A common motivation for using spatial microsimulation is lack of data on a specific variable (as with the CakeMap example in the previous chapter). In such cases internal validation, combined with proxy variables for which external datasets are available, may be the best approach to model evaluation. This is the case with the CakeMap example explored in the previous chapter. There are no readily available datasets on the geographic distribution of cake consumption, so external validation of the dependent variable (frequency with which cake is eaten) is deemed impossible in this case. However, new sources of data such as number of confectionery shops, consumer surveys and even social media could be explored to provide 'sanity checks' on the results. Sometimes you may need to be creative to find data for external validation.

What is important to note on these two kinds of validation is that they test the model at different levels. *Internal validation* tests the model quality at the aggregate level, assuming the input data is relevant to the research question, accurate and representative. If this validation fails, you may have a problem with the input microdata (e.g. an excess of 'empty cells' or an unrepresentative sample of the population), implementation of the population synthesis algorithm, or contradictory constraints. Internal validation highlights problems of method: if internal validation results are poor, the cause of the problem should be diagnosed (e.g. is it poor data or poor implementation?) and fixed.

By contrast, *external validation* compares the model results with data that is *external* to the model. External validation is more rigorous as it relates simultaneously to the model's performance and whether the input data are suitable for answering the research questions explored by spatial microsimulation. Poor external validation results can come from everywhere, so are harder to fix (internal validation can rule out faulty methods, however). Thus internal and external validation complement each other.

This chapter explains how to undertake routine checks on spatial microsimulation procedures, how to identify outlying variables and zones which are simply not performing well (internal validation) and how to undertake external validation. Even in cases where there is a paucity of data on the target variable, as with cake consumption, there is usually at least some tests of the model's performance against external data that can be undertaken. As we will see with the CakeMap example (where income is used as a proxy variable for external validation purposes) this can involve the creation of new target variables, purely for the purposes of validation.

8.1 Internal validation

Internal validation is the process of comparing the model's output against data that is internal to the model itself. In practice this means converting the synthetic spatial microdata into a form that is commensurate with the constraint variables and comparing the two geographically aggregated datasets: the observed vs simulated values. Every spatial microsimulation model will have access to the data needed for this comparison. Internal validation should therefore be seen as **the bare minimum** in terms of model evaluation, to be conducted as a standard procedure on all spatial microsimulation runs. When authors refer to this procedure as "result validation" they are being misleading. Internal validation tells us simply that the results are internally consistent; it should always be conducted. The two main causes of poor model fit in terms of internal validation are:

1. Incorrectly specified constraint variables. For example the total number of people in each zone according to one variable (e.g. age) may be different from that according to another (e.g. employment status). This could be because each variable uses a different *population base* (see Glossary).

2. *Empty cells*. These represent 'missing people' in the input microdata who have a combination of variables that are needed for good model fit. If in the input microdata for SimpleWorld there were no older males, for example, the model would clearly perform much worse.

Other sources of poor fit between simulated and observed frequencies for categories in the linking variables include simple mistakes in the code defining the model, incorrect use of available algorithms for population synthesis and, to a lesser extent, integerisation (Lovelace et al. 2015).

Because internal validation is so widely used in the literature there are a number of established measures of internal fit that have been used. Yet there is little consistency in the measures that are used. This makes it difficult to assess which models are performing best across different studies, a major problem in spatial microsimulation research. If one study reports only r values, whereas another reports only *TAE* (each measure will be described shortly), there is no way to assess which is performing better. There is a need for more consistency in reporting internal validation. Hopefully this chapter, which provides descriptions of each of the commonly used and recommended measures of *goodness-of-fit* as well as guidance on which to use — is a step in the right direction.

Several metrics of model fit exist. We will look at some commonly used measures and define them mathematically before coding them in R and, in the subsequent section, implementing them to evaluate the CakeMap spatial microsimulation model. The measures developed in the section are:

- Pearson's correlation (r), a formula to quantify the linear correlation between the observed and final counts in each of the categories for every zone.
- Total absolute error (TAE), also know as the sum of absolute error (SAE), simply the sum of absolute (positive) differences between the observed and final counts in each of the categories for every zone.
- Relative error (RE), the TAE divided by the total concerned population.
- Mean relative error (MRE), the sum of RE calculated per category and per zone.
- Root mean squared error (RMSE), the square root of the mean of all squared errors. This metric emphasises the relative importance of a few large errors over the build-up of many small errors.
- Chi-squared, a statistical hypothesis test that compares two hypotheses and determines which one is more probable. First hypothesis is that the final and the observed counts follow the same distribution. Second hypothesis is the opposite.

Often, to illustrate the internal quality of the model, we add some representations. Those can be maps or graphs. We will proceed to some representation for the example of CakeMap.

8.1.1   Pearson's r

Pearson's coefficient of correlation (r) is the most commonly used measure of aggregate level model fit for internal validation. r is popular because it provides a fast and simple insight into the fit between the simulated data and the constraints at an aggregate level. In most cases r values greater than 0.9 should be sought in spatial microsimulation and in many cases r values exceeding 0.99 are possible, even after integerisation.

r is a measure of the linear correlation between two vectors or matrices. In spatial microsimulation, if the model works, the observed and final counts in each of the categories for every zone are equal. This means that, when plotting, for each category, the observed counts versus the final counts, we have a perfect line where abscissa and ordinates are equal. Thus, the measure of a *linear* correlation is the one needed. The formula to calculate the Pearson's correlation between the vectors (or matrices) x and y is:

$$r = \frac{s_{XY}}{S_X S_Y} = \frac{\frac{1}{n}\sum_{i=1}^{n} x_i y_i - \bar{x}\bar{y}}{\sqrt{\frac{1}{n}\sum_{i=1}^{n} x_i^2 - \bar{x}^2}\sqrt{\frac{1}{n}\sum_{i=1}^{n} y_i^2 - \bar{y}^2}}$$

This corresponds to the covariance divided by the product of the standard deviation of each vector. This can sound complicated, but it is just a standardized covariance. If the fit is perfect, both vectors (simulated and constraint) have the same values and the covariance is equal to the product of the standard deviations. Thus, the r is then close to 1.

Note that this measure is very influenced by outliers in the vectors. This means that if only one category has a bad fit, the r value is very affected.

8.1.2 Absolute error measures

TAE and RE are crude yet effective measures of overall model fit. TAE has the additional advantage of being very easily understood as simply the sum of errors:

$$e_{ij} = obs_{ij} - sim_{ij}$$

$$TAE = \sum_{ij} |e_{ij}|$$

where e is error, *obs* and *sim* are the observed and simulated values for each constraint category (j) and each area (i), respectively. Note the vertical lines | mean we take the absolute value of the error. This means that an error of -5 has the same impact as an error of +5. This avoids the possibility of having an error of 0 if for one category **obs** is bigger and for another **obs** is smaller. It really counts the number of differences.

RE is the TAE divided by the total population of the study area. This means that if we compare the results for all variables simultaneously, we divide the TAE by the total population multiplied by the number of variables[2]. Thus, TAE

---

[2]This is very intuitive, since by considering all constraint variables together, we have taken the whole population once for each variable.

is sensitive to the number of people and of categories within the model, while RE is not. RE can be interpreted as the percentage of error and corresponds just to the standardised version of TAE.

$$RE = TAE/(total\_pop * n\_var)$$

Mean relative error (MRE) needs first to calculate RE per variable and per zone.

$$MRE = \sum_{i=1}^{n\_var} \sum_{j=1}^{n\_zones} RE(zone = j; var = j)$$

Before seeing how these metrics can easily be implemented in code, we will define the other metrics defined in the above bullet points. Of the three 'absolute error' measures, we recommend reporting RE or MRE, as it scales with the population of the study area.

8.1.3 Root mean squared error

RMSE is similar to the absolute error metrics, but uses the *sum of squares* of the error. Recent work suggests that RMSE is preferable to absolute measures of error when the errors approximate a normal distribution (Chai and Draxler, 2014). Errors in spatial microsimulation tend to have a 'normal-ish' distribution, with many very small errors around the mean of zero and comparatively few larger errors. RMSE is defined as follows:

$$RMSE = \sqrt{\frac{1}{n}\sum_{i}^{n} e_i^2}$$

RMSE is an interesting measure of the error, since TAE and RE would be the same if the errors are $(1, 1, 1, 1)$ or $(0, 0, 0, 4)$. However, we consider the fit as globally better if it contains several few errors than if it is perfect for 3 zones and a higher error for the fourth. In this case, RMSE will detect this difference. For the first case, RMSE is $\sqrt{\frac{4}{4}}$. For the second case, RMSE equals $\sqrt{\frac{4^2}{4}} = 2$.

As with TAE, there is also a standardised version of RMSE, normalised root mean error squared (NRMSE). This is calculated by dividing RMSE by the range of the observed values:

$$NRMSE = \frac{RMSE}{max(obs) - min(obs)}$$

8.1.4 Chi-squared

Chi-squared is a commonly used test of the fit between absolute counts of categorical variables. It has the advantage of providing a *p value*, which represents the chances of obtaining a fit between observed and simulated values through chance alone. It is primarily used to test for relationships between categorical variables (e.g. socio-economic class and smoking) but has been used frequently in the spatial microsimulation literature (Voas and Williamson 2001; Wu et al. 2008).

The chi-squared statistic is defined as the sum of the square of the errors divided by the observed values (Diez et al., 2012). Suppose we have a simulated matrix *sim* (for example, simulated counts) and an observed matrix *obs*, the chi-squared is calculated by:

$$\chi^2 = \sum_{i=1}^{n\_line} \sum_{j=1}^{n\_column} \frac{(sim_{ij} - obs_{ij})^2}{obs_{ij}}$$

The *chi-squared* test is the probability of obtaining the calculated χ^2 value or a worst (in terms of validating models, worst means a bigger difference), given the number of *degrees of freedom* (representing the number of categories) in the test.

An advantage of chi-squared is that it can compare vectors as well as matrices. As with all metrics presented in this section, it can also calculate fit for subsets of the data. A disadvantage is that chi-squared does not perform well when expected counts for cells are below 5. If this is the case it is recommended to use a subset of the aggregate level data for the test (Diez et al., 2012).

8.1.5 Which test to use?

The aforementioned tests are just some of the most commonly used and most useful *goodness of fit* measures for internal validation in spatial microsimulation. The differences between different measures are quite subtle. Voas and Williamson (2001) investigated the matter and found no consensus on the measures that are appropriate for different situations. Ten years later, we are no nearer consensus.

Such measures, that compare aggregate count datasets, are *not* sufficient to ensure that the results of spatial microsimulation are reliable; they are methods of *internal validation*. They simply show that the individual level dataset has been reweighted to fit with a handful of constraint variables: i.e. that the process has worked under its own terms.

Our view is that all the measures outlined above are useful and roughly analogous (a perfect fit will mean that measures of error evaluate to zero and that $r = 1$). However, some are better than others. Following Chai and Draxler (2014), we recommend using r as a simple test of fit and reporting *RMSE*, as it is a standard test used across the sciences. *RMSE* is robust to the number of observations and, using *NRMSE*, to the average size of zones also. Chi-squared is also a good option as it is very mature, provides *p values* and is well known. However, chi-squared is a more complex measure of fit and does not perform well when the table contains cells with less than 5 observations, as will be common in spatial microsimulation models of small areas and many constraint categories.

We recommend reporting more than one metric, while focusing on measures that you and your colleagues understand well. Comparing the results with one or more alternative measures will add robustness. However, a more important issue is external validation: how well our individual level results correspond with the real world.

8.1.6   Internal validation of CakeMap

Following the 'learning by doing' ethic, let us now implement what we have learned about internal validation. As a very basic test, we will calculate the correlation between the constraint table cells and the corresponding simulated cell values for the CakeMap example:[3]

```
cor(as.numeric(cons), as.numeric(ind_agg))
```

```
## [1] 0.9968529
```

We have just calculated our first goodness-of-fit measure for a spatial microsimulation model and the results are encouraging. The high correlation suggests that the model is working: it has internal consistency and could be described as 'internally valid'. Note that we have calculated the correlation before integerisation here. In the perfect fit, we would have a linear correlation of exactly 1.

In micro-simulation, we have the whole population with all characteristics of each individual, only after the simulation. For this reason, we have to aggregate the simulated population to have a matrix comparable with the constraint. In this sense, there are two ways to proceed. First, we can make the comparison variable per variable and the total number of individuals is the constraint number of people in the area. Secondly, we can take all variables together,

---

[3]Data frames will not work in this function and must be converted to matrices with `as.numeric`.

meaning having a matrix including the whole population for each variable. This implies that the sum of all cells equals to the multiplication of the number of people in the area by the number of variables. Our choice here is the second alternative. Then, if we need more details on the fit in one zone, we can proceed to an analysis per variable for this specific case.

We can also calculate the correlation of these two vectors zone per zone. By this way, we will be able to notify for which zones our simulation could be less representative. A vector of the correlation per zone, called `CorVec`, is calculated:

```
# initialize the vector of correlations
CorVec <- rep (0, dim(cons)[1])

# calculate the correlation for each zone
for (i in 1:dim(cons)[1]){
  num_cons <- as.numeric(cons[i,])
  num_ind_agg <- as.numeric(ind_agg[i,])
  CorVec[i] <- cor (num_cons, num_ind_agg)
}
```

We can then proceed to a statistical analysis of the correlations and identify the worst zone. In the code below, the summary of the vector of correlation is performed. The minimum value is 0.9451. This is the performance of the zone 84. This value is under the global correlation, but still close to 1. We can also observe that the first quartile is already 1. This means that for more than 75% of the zones, the correlation is perfect (at least with an approximation to 4 decimals). Moreover, by identifying the second worst zone, we can see that its correlation is around 0.9816.

```
# summary of the correlations per zone
summary (CorVec)
```

```
##    Min. 1st Qu.  Median    Mean 3rd Qu.    Max.
## 0.9451  1.0000  1.0000  0.9993  1.0000  1.0000
```

```
# Identify the zone with the worst fit
which.min(CorVec)
```

```
## [1] 84
```

```
# Top 3 worst values
head(order(CorVec), n = 3)
```

```
## [1] 84 82  7
```

This ends our analysis of correlation. Next we can calculate total absolute error (TAE), which is easily defined as a function in R:

```
tae <- function(observed, simulated){
  obs_vec <- as.numeric(observed)
  sim_vec <- as.numeric(simulated)
  sum(abs(obs_vec - sim_vec))
}
```

By applying this function to CakeMap, we find a TAE of 26445.57, as calculated below. This may sound very big, but remember that this measure is very dependent on the scale of the problem. 26,445 may seem like a large number but it is small compared with the total population multiplied by the number of constraints: 4,871,397. For this reason, the relative error RE (also called the standardised absolute error) is often preferable. We observe a RE of 0.54%. Note that RE is simply TAE divided by the total of all observed cell values (that is, the total population of the study area multiplied by the number of constraints).

```
# Calculate TAE
tae(cons, ind_agg)
```

```
## [1] 26445.57
```

```
# Total population (constraint)
sum(cons)
```

```
## [1] 4871397
```

```
# RE
tae(cons, ind_agg) / sum(cons)
```

```
## [1] 0.005428745
```

As with all tests of goodness of fit, we can perform the analyses zone per zone. For the example, we call the vector of TAE and RE per zone, respectively, TAEVec and REVec.

```
# Initialize the vectors
TAEVec <- rep(0, nrow(cons))
REVec <- rep(0, nrow(cons))

# calculate the correlation for each zone
for (i in 1:nrow(cons)){
  TAEVec[i] <- tae (cons[i,], ind_agg[i,])
  REVec[i] <- TAEVec[i] / sum(cons[i,])
}
```

The next step is to interpret these results. The summary of each vector will help us. Note that in the best case, the correlation is high, but the RE and TAE are small. The zone with the highest error is also the number 84, which has a TAE of 14710 individuals times variables and a RE of 21.3%. This zone seems to have a simulation a bit distant from the constraint. By watching the second and third worst zone, we can see that its RE is respectively around 12.5% and 7.0%. The third quartile is of order 10^{-5} (10^{-3}%). This is pretty close to 0. Thus, 75% of the zones has a RE smaller than the third quartile. The maximum values aside, it appears that for the majority of the zones, the RE is small.

```
# Summary of the TAE per zone
summary (TAEVec)
```

```
##    Min. 1st Qu. Median   Mean 3rd Qu.    Max.
##     0.0     0.0    2.0  213.3     2.0 14710.0
```

```
# Summary of the RE per zone
summary (REVec)
```

```
##       Min.   1st Qu.    Median      Mean   3rd Qu.      Max.
## 0.000e+00 0.000e+00 4.694e-05 3.379e-03 6.200e-05 2.132e-01
```

```
# Identify the worst zone
which.max(TAEVec)
```

```
## [1] 84
```

```
which.max(REVec)
```

```
## [1] 84
```

```
# Maximal value
tail(order(TAEVec), n = 3)
```

```
## [1]   7 82 84
```

```
tail(order(REVec), n = 3)
```

```
## [1]   7 82 84
```

Similar analyses can be applied for the other tests of goodness of fit. In all cases, it is very important to have an idea of the internal validation of your model. For example, if we want to analyse the cake consumption by using your synthetic population created here, we have to be aware that for the zone 84, the model does not fit so well the constraints.

Knowing that zone 84 is problematic, the next stage is to ask "how problematic?". If a single zone is responsible for the majority of error, this would suggest that action needs to be taken (e.g. by removing the offending zone or by identifying which variable is causing the error).

To answer the previous question numerically, we can rephrase it in technical terms: "Which proportion of error in the model arises from the worst zone?" This is a question we can answer with a simple R query:

```
worst_zone <- which.max(TAEVec)
TAEVec[worst_zone] / sum(TAEVec)
```

```
## [1] 0.5561611
```

The result of the above code demonstrates that more than half (56%) of the error originates from a single zone: 84. Therefore zone 84 certainly is anomalous and worthy of further investigation. An early strategy to characterise this zone and compare it to the others is to visualise it.

To this end, Figure 8.1 places the TAE values calculated previously on a map, with a base-layer supplied by Google for context — see the book's online source code (https://github.com/Robinlovelace/spatial-microsim-book/blob/master/validation.Rmd) to see how. Zone 84 is clearly visible in this map as a ward just above Leeds city centre. This does not immediately solve the problem, but it confirms that only few zones have bigger errors.

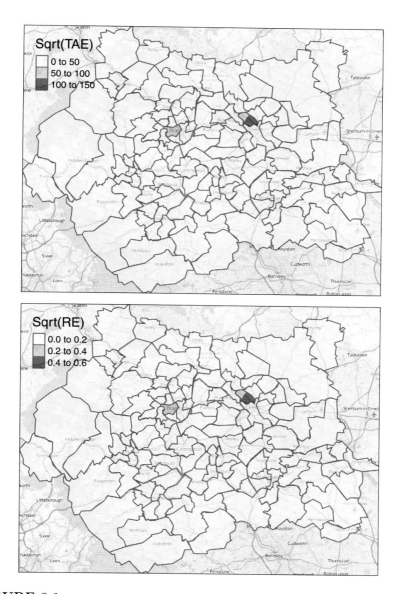

FIGURE 8.1

Geographical distribution of Total Absolute Error (TAE) and Relative Error (RE). Note the zones of high error are clustered in university areas such as near the University of Leeds, where there is a high non-resident population.

Note that the maps presented in Figure 8.1 look identical for TAE and RE values except for the scale; the similitude of these measures of fit can be verified using a simple correlation:

```
cor(TAEVec, REVec) # the measures are highly correlated
```

```
## [1] 0.9963713
```

In this case, both are quite well correlated. However, when having very different zones, in terms of total population, it can have more differences between the two maps. Indeed, with the same TAE, if the zone 84 had contained a total population two times bigger, the RE would be very smaller. Thus, RE would be divided by the multiplication of 2 and the number of variables.

Having identified a zone that is particularly problematic (the 84), we will look at the responsible variables. We focus on the zone 84 and calculate the number of differences between the constraint and the simulation for each category:

```
RudeDiff <- cons[84,] - ind_agg[84,] # differences for zone 84
diff <- round( abs(RudeDiff) ) # interesting differences
```

```
diff[diff > 1000] # printing the differences bigger than 1000
```

```
## m45_54 m55_64    Car  NoCar
##   1274   1120   4960   4960
```

The responsible variable seems to be the car ownership. We have made a similar check for the three worst zones and this variable is always the one with the largest difference. To investigate the reasons for this, we print the constraints for this variable inside the three worst zones and the marginals of the observed individuals:

```
worst <- tail(order(REVec), n = 3)
cons[worst, c("Car", "NoCar")] # constraint for 3 worst zones
```

```
##          Car NoCar
## [1,]   6983 10927
## [2,]  11453  8040
## [3,]   8796 14204
```

```
table( ind[,2] ) # individuals to weight (1 = Car ; 2= NoCar)
```

```
##
##   1   2
## 738 178
```

Only few observed individuals did not own a car. Thus, for zones needing a lot of persons that have no car, the weight of only 178 individuals out of 916 can be adapted. The possibility of having an individual that has the whole range of possible characteristics is then lower. The individuals without a car are saved in the NoCar variable. The contingency table of these people for the number of cakes and the age shows that we have nobody of age 55-64 eating more than 6 cakes.

```
# individuals not owning a car
NoCar <- ind[ind$Car==2,]

# Cross table
table(NoCar$NCakes,NoCar$ageband4)
```

```
##
##         16-24 25-34 35-44 45-54 55-64 65-74
## <1          3     5     3     5     4     6
## 1-2        10    11     8     5     5     5
## 3-5         9    13     6    10     3     8
## 6+          7    10     3     4     0    10
## rarely      2     7     2     6     2     6
```

The three zones with the worst simulation needed a lot of people without a car. On the contrary, below, we print the constraint of car of the three best zones. They needed less people of this category. This is the risk by generating a population of 1,623,800 of inhabitants and having a survey including only 916 persons.

```
best <- head(order(REVec), n = 3)
 # constraint for 3 best zones
cons[best, c("Car", "NoCar")]
```

```
##          Car NoCar
## [1,] 11478  2853
## [2,] 13958  2576
## [3,]  9449  1896
```

In conclusion, the simulation runs well for all zone excepts few ones. This is due to the individuals present in the sample. This could be explained by a survey that was not uniformly distributed through the different zones or does not include enough persons.

8.2   Empty cells

Roughly speaking, 'empty cells' refer to individuals who are absent from the input microdata. More specifically, empty cells represent individuals with a combination of attributes in the constraint variables that are likely to be present in the real spatial microdata but are known not to exist in the individual level data. Empty cells are easiest to envisage when the 'seed' is represented as a contingency table. Imagine, for example, if the microdata from SimpleWorld contained no young males. The associated individual data could be Table 8.1, leading to the cross table shown in Table 8.2. We can clearly identify that there is no young male. Applying reweighting methods on this kind of input data result in an unrealistic final population, without young male.

	age	sex
1	a50+	m
2	a50+	m
4	a50+	f
5	a0_49	f

TABLE 8.1: Individual level data from SimpleWorld with empty cells. Note there are no young males.

	f	m
a0_49	1	0
a50+	1	2

TABLE 8.2: Contingency table of the SimpleWorld microdata with no young males. Note the zero: this is the empty cell.

The importance of empty cells and methods for identifying whether or not they exist in the individual level is explained in a recent paper (Lovelace et al. (2015)). The number of different constraint variable permutations ($Nperm$) increases rapidly with the number of constraints (see equation 8.1 below), where $n.cons$ is the total number of constraints and $n.cat_i$ is the number of categories within constraint i:

$$Nperm = \prod_{i=1}^{n.cons} n.cat_i \qquad (8.1)$$

To exemplify this equation, the number of permutations of constraints in the SimpleWorld microdata example is 4: 2 categories in the sex variables multiplied by 2 categories in the age variable. Clearly, $Nperm$ depends on how continuous variables are binned, the number of constraints and diversity within each constraint. Once we know the number of unique individuals (in terms of the constraint variables) in the survey ($Nuniq$), the test to check a dataset for empty cells is straightforward, based on equation 8.1:

$$is.complete = \left\{ \begin{array}{ll} TRUE & \text{if } Nuniq = Nperm \\ FALSE & \text{if } Nuniq < Nperm \end{array} \right\} \qquad (8.2)$$

Once the presence of empty cells is determined, the next stage is to identify which types of individuals are missing from the individual level input dataset (Ind).

The 'missing' individuals, needed to be added to make Ind complete, can be defined by the following equation :

$$Ind_{missing} = \{x | x \in complete \cap x \notin Ind\}$$

This means simply that the missing cells are defined as individuals with constraint categories that are present in the complete dataset but absent from the input data.

8.3   External validation

Beyond mistakes in the code, more fundamental questions should be asked of results based on spatial microsimulation. The validity of the assumptions affect confidence one should have in the results. For this we need external datasets. External validation is therefore a tricky topic which is highly dependent on the available data (Edwards et al., 2010).

Geocoded survey data, *real* spatial microdata, is the 'gold standard' when it comes to official data. Small representative samples of the population for small areas can be used as a basis for individual level validation.

8.4   Chapter summary

This chapter explored methods for checking the results of spatial microsimulation, building on the CakeMap example presented in the previous chapter. This primarily involved checking that the results are internally consistent: that the output spatial microdata correspond with the geographical constraints. This process, generally referred to as 'internal validation', is important because it ensures that the model is internally consistent and contains no obvious errors.

However, the term 'validation' can be misleading as it implies that the model is in some way 'valid'. A model is only as good as its underlying assumptions, which may involve some degree of subjectivity. We therefore advocate talking about this phase as 'evaluation' or simply 'model checking', if all we are doing is internal validation.

In the example of CakeMap, no datasets are available to check if the simulated rate of cake is comparable with that estimated from other sources. In the case of microsimulation, external validation is often difficult because available datasets are usually used for the simulation. This helps explain why internal validation is far more common in spatial microsimulation studies than external validation, although the latter is generally more important.

9

Population synthesis without microdata

CONTENTS

Sometimes no representative individual level dataset is available as an input for population synthesis. In this case, the methods described in the previous chapters must be adapted accordingly. The challenge is still to generate spatial microdata that fits all the constraint tables, but based on a purely synthetic 'seed' input cross-tabulated contingency table. Many combinations of individual level data could correspond to these distributions. Depending on the aim of the spatial microsimulation model, simply finding one reasonable fit can be sufficient.

In other cases a fit based on *entropy maximisation* may be required. This concept involves finding the population that is most likely to represent the micro level populations (Bierlaire, 1991) (M., 1991). This chapter demonstrates two options for population synthesis when real individual level data is unavailable:

- *Global cross-tables and local marginal distributions* (Section 9.1) explains a method for cases where the constraints consist in cross-tables not spatially located and local marginal distributions.
- *Two level aggregated data* (Section 9.2) contains a procedure to make a spatial microsimulation when having data at different aggregated levels, for example, one for the provinces and one for the districts.

9.1 Global cross-tables and local marginal distributions

Assume we have a contingency table of constraint variables for the entire study area (but not at the local level) in the aggregate level data. This multi-dimensional cross-table (the seed) could be the result of a previous step such as the implementation of IPF re-weight individual level data to fit the case-study area of interest.

If the marginal distributions for small areas are known, we can use the **mipfp** function as previously shown. If, however, the only information about the zones is the total population living there, the function is usable only when considering the zone as a variable. In this specific case, having no additional data, the only option corresponds to re-scale the global cross-table for each zone. Note that this implies that the correlations between the variables are independent of the zone in question.

To illustrate, we will develop the SimpleWorld example (which can be loaded from the book's data directory by entering `source("R/SimpleWorld.R")` or was previously loaded if you as followed (Chapter 4)) with adapted constraints. When watching the available data in an aggregated level, we have for the moment:

```
# Cross-tabulation of individual level dataset
table(ind$age, ind$sex)
```

```
##
##         f m
##   a0_49 1 1
##   a50+  1 2
```

```
(total_pop <- rowSums(con_sex)) # total population of each zone
```

```
## [1] 12 10 11
```

To illustrate this section, the local constraint will be the total number of people in each zone (last column of `consTot`). The global constraint is a matrix of the form of the cross-table between age and sex, but including the total population (33 people for SimpleWorld). The new constraints could be:

```
# Global Constraint possible for SimpleWorld
global_cons <- table(ind$age, ind$sex)
global_cons[1,] <- c(6,9)
global_cons[2,] <- c(7,11)

# Local Constraint for SimpleWorld
local_cons <- total_pop
```

When only the total population is known for each zone, the best way to create a synthetic population is to simply re-scale the cross-table. For each zone, a table proportional to the global one is created. The results are stored in a three dimensional array, which first dimension represents the zone. The initialisation of the resulting matrix is the first step. We here fill in the table with "0".

```
# initialise result's array and its names
resNames <- list(1:nrow(cons), rownames(global_cons),
          colnames(global_cons))
res <- array(0, dim=c(nrow(cons), dim(global_cons)),
          dimnames=resNames)
```

Now the final weight table is calculated, simply by taking the global matrix and re-scaling it to fit the the desired marginals. In this way we keep the global proportions, but with the correct total per zone. Note that making this process is exactly the same as running `mipfp` on the seed table with as constraints only the zone marginals.

```
# Re-scale the cross-table to fit the zone's constraints
for (zone in 1:length(total_pop)){ # loop over the zones
  res[zone,,] <- global_cons * total_pop[zone] / sum(global_cons)
}

# Print the cross-table for zone 1
res[1,,]
```

```
##          f    m
## a0_49 2.18 3.27
## a50+  2.55 4.00
```

We can verify that the total population per zone is of the desired size. We can also check the global table of age and sex. This means that we have now weights fitting well all available data.

```
# Check the local constraints for each zone (should be TRUE)
for (zone in 1:length(total_pop)){
 print( sum(round(res[zone,,])) == total_pop[zone] )
}
```

```
## [1] TRUE
## [1] TRUE
## [1] TRUE
```

```
# Save the global final table
SimTot <- apply(res,c(2,3),sum)

# Check the global constraint (should be 0)
sum(SimTot - global_cons)
```

```
## [1] 0
```

As with IPF, the fractional result needs to be integerised to create spatial microdata. The `round()` function generally provides a reasonable approximation, in terms of fitting the constraints. However, the aforementioned integerisation algorithms such as *truncate, replicate, sample* (TRS) can also be used. This method cannot be followed exactly, because we want to perfectly fit the few constraints we have. In our example, a satisfactory result is achieved by using the round function, as shown in the code below.

```
# Integerisation by simply using round
resRound <- round(res)
resTruncate <- floor(res) # take the minimum integer value

# Zero error achieved by rounding for global constraint
apply(resRound, c(2,3), sum) - global_cons
```

```
##        f m
## a0_49 0 0
## a50+  0 0
```

```
# Zero error achieved by rounding for local constraint
apply(resRound,c(1),sum) - local_cons
```

```
## 1 2 3
## 0 0 0
```

It is due to luck (and the small size of the SimpleWorld example) that the `round` method works in this case: in most cases there will be errors due to rounding. If a zone had 4 individuals and three categories, for example, the resulting weights could be $(\frac{4}{3}, \frac{4}{3}, \frac{4}{3})$. Then, the rounding gives $(1, 1, 1)$ and there would be too few individuals in the synthetic population (3 not 4). We can try the algorithms proposed in (Section 5.3). However, as illustrated by the following code chunks, these integerisation methods lead to errors in relation to the constraints.

```
# Integerisation with pp
res_pp <- int_pp(res)

apply(res_pp, c(2,3), sum) - global_cons
```

```
##
##       f  m
##  a0_49  0  3
##  a50+   1 -4
```

These errors are often very small and if you model a whole country, the relative error is small. Note that this little error comes from the random draw at last stage of the algorithm. Here, TRS is better than PP, as explained in Chapter 5.

```
# Integerisation with trs
set.seed(17)
res_trs <- res_pp <- array(dim = dim(res))
# Apply trs (see R/functions.R to see how int_trs works)
res_trs[] <- int_trs(res)

# Print the errors
apply(res_trs, c(2,3), sum) - global_cons
```

```
##
##       f  m
##  a0_49  0 -1
##  a50+   1  0
```

If desired, we can adapt TRS to ensure it fits fit the constraints at the end of the process. To adapt the method to use TRS, we first truncate[1] the data and identify the missing individuals, in terms of constraints.

```
# Truncate
resTruncate <- floor(res)

# number of missing individuals
sum(total_pop) - sum(resTruncate)
```

```
## [1] 4
```

---

[1] Note that truncate means round each weight to the first integer under the weight. This implies that we underestimated the population.

This means that, in total, 4 individuals are missing after we have truncated. We will have to chose which categories will be incremented. For this, the basic TRS take the decimal parts of the weights (that were forgotten when truncate) and make a random draw inside this distribution. This is in this step that we can add an error in terms of the constraints. To make a better fit after integerisation, we need to observe in which category and in which zone we have to add individuals.

```
# Calculate the total simulated cross-table
# After truncate
SimTotTruncate <- apply(resTruncate,c(2,3),sum)

# Number of missing individuals per category
# After truncate
ToAdd <- global_cons - SimTotTruncate
ToAdd
```

```
##
##         f m
##   a0_49 1 1
##   a50+  1 1
```

```
# Number of missing individuals per zone
# After truncate
ToComplete <- local_cons - apply(resTruncate,c(1),sum)
ToComplete
```

```
## 1 2 3
## 1 2 1
```

We observe that the individuals to add are one per category of age and sex. In terms of zones, one individual is missing in zone one and three, whereas two persons will have to be added in zone 2.

The principle now is to add people in the not completed zones and categories. The cells to be incremented are always chosen as the one with the bigger decimal parts (whereas in TRS, these decimals act as probabilities). Note that we chose to adapt the `resTruncate` instead of defining another tabular.

The code works as followed: As long as there are missing individuals in the matrix, we look at next biggest decimal and add an individual in the corresponding cell if someone is missing in this category and in this zone.

```
# Calculate the decimals left by truncate
decimals <- res - resTruncate

# Adapting resTruncate to fit all constraints
while (sum(total_pop) - sum(resTruncate) > 0){
  # find the biggest decimals
  i <- which( decimals == max(decimals), arr.ind = TRUE)

  # remember we already considered this cell
  decimals[i] <- 0

  # if this zone still miss individuals
  if (ToComplete[i[1]] > 0){
    # if this category still miss individuals
    if (ToAdd[i[2],i[3]] > 0){
      resTruncate[i] <- resTruncate[i] + 1
      ToComplete[i[1]] <- ToComplete[i[1]] - 1
      ToAdd[i[2],i[3]] <- ToAdd[i[2],i[3]] - 1
    }
  }
}
```

The new values in `resTruncate` follow all constraints. The adaptation of TRS could be avoided by using combinatorial optimization to integerise. Indeed, this process could choose for each cell to take the integer just under or just above the weight by optimising the fit to the constraints. We use TRS here because it is faster and requires fewer lines of code (see Lovelace and Ballas, 2013 for more detail).

After the integerisation, the last step to get the final individual dataset is the expansion. This stage is intuitive, since we have now a table containing the number of individuals in each category. Thus, we simply need to replicate the combination of categories the right number of times.

We can first flatten the 3–dimensional matrix. Then, the final individual micro dataset `ind_data` is created.

```
countData <- as.data.frame.table(resTruncate)
indices <- rep(1:nrow(countData), countData$Freq)
ind_data <- countData[indices,]
```

9.2   Two level aggregated data

We present here how to find a possible distribution per zone when having only aggregated data, but in two different levels of aggregation. For example, we have some data for the municipalities and other for the districts. A first proposition can be to use a genetic algorithm that minimises the distance between the constraint and the simulation. This can give very good solutions, but need a high level understand of optimisation and is rare in the literature for the moment. For classical problems, a simpler method is available. The basis of this method is explained here. The solution proposed by Barthélemy and Toint (2013), and used in this book, is to generate a 'seed' before executing IPF.

This paper demonstrates the simulation of a population with four characteristics per individual: the gender, the age class, the diploma level and the activity status and at the municipality level. Their available data was:

1. At municipality level: the cross table gender x age and the marginals of diploma level and activity status;
2. At district level: the cross tables gender x activity status, gender x diploma level, age x activity status and age x diploma level.

Note that a district contains several municipalities, but each municipality is associated to only one district. We consider the marginals of the tables being consistent. If not, a preliminary step is necessary to re-scale the data to avoid shifting to probabilities. We chose to do this to have the best chance to fit the data well. When shifting to probabilities, it is more difficult to adapt the distributions during the iteration. Indeed, when considering the theoretical counts, if you create a young women, you just need to take the cell 'young' and 'women' and make minus one. When considering probabilities, when you create a young woman, you have to recalculate all probabilities, because you still need less women and proportionally, more men than before. This is the reason why we prefer to adapt the distributions to the one we are the more confident.

The global idea of their method is to proceed in two steps. First, they simulate the cross table of the four variables per district. Then, this table is considered as the seed of the IPF algorithm to simulate the distributions per municipality. During this second stage, the data concerning the municipality are used as constraints. How to execute the second part has been explained in the first section of this chapter. The point here is to develop the process, per district, to simulate the four–dimensional cross table, with the available data. This is also done in two steps :

1. Generate age x gender x diploma level and age x gender x professional status;
2. Generate age x gender x diploma level x professional status.

For the first step, we will explain only the creation of the first cross table, since the second reasoning is similar. The idea is simply to proceed proportionally to respect both available tables. The pseudo-code below corresponds to the code provided by Barthélemy and Toint (2013).

For the clarity of the formal formula, we rename gender (A), age (B) and diploma level (C). To create the cross table of these three variables, we have at the district level the cross tables gender x diploma level (renamed AC) and age x diploma level (renamed BC). Then, the cells of the three–dimensional table is defined for each gender g, age a and diploma level d as followed :

$$ABC(g, a, d) = \frac{AC(g, d)}{margin(d)} BC(a, d)$$

The formula is intuitive. The fraction gives the proportion of each gender inside the considered category of diploma level. Then, this proportion splits the number of persons having characteristics a and d into the category of g. For example, in the specific case of defining (Male, Young, Academics), we will have :

$$ABC(Male, Young, Aca) = \frac{AC(Male, Aca)}{\#Aca} BC(Young, Aca)$$

Suppose we have 50 young academics out of 150 academics (90 female and 60 male). We would have:

$$ABC(Male, Young, Aca) = \frac{60}{150} * 50 = 20$$
$$ABC(Female, Young, Aca) = \frac{90}{150} * 50 = 30$$

Thus, the tables age x gender x diploma level and age x gender x professional status are simulated. The seed for the IPF function can now be established, with help of the two contingencies. These initial weights will be the distribution of the four variables inside the whole district.

This seed is generated by several iterations. The initialisation of the cross table is simply a matrix with the right number of dimensions, with "0" in impossible cells and "1" in potentially non empty cases. For example, individuals of less than 10 years cannot hold a diploma from university.

With this initial point, an IPF can be performed to fit the two previously determined three–dimensional tables. The result is a table with the four variables per district.

The final step is explained in the previous section. Indeed, we have a contingency table at the district level and the zone margins. Note that you can imagine a lot of combinations of IPF to perform a population synthesis adapted to your own data.

9.3   Chapter summary

In summary spatial microsimulation can be used in situations where no sample data is available. The techniques can be adapted to work with a synthetic population. This chapter presented two methods for creating synthetic populations in R, the selection of which should depend on the type of input constraint data available. The first method assumed access to global cross tables and local marginals. The second assumed having aggregate data at different levels.

10

Household allocation

CONTENTS

So far, this book has explored data on 2 levels: the individual level and the level of administrative zones. The household is another fundamental building block of human organisation around which key decision-making, economic and data-collecting activities are centred. We will here develop results for Belgium in a specific study. Note that this chapter is based on a research made by Dumont Morgane (UNamur) and funded by the Wallonia Region of Belgium. Timoteo Carletti (UNamur), Eric Cornélis (UNamur), Philippe Toint (UNamur) and Thierry Eggericks (UCL Louvain-La-Neuve) were involved in the research. The academic groups of DEMO from UCL-Louvain-La-Neuve and the OWS (Observatoire Wallon de la Santé) also provided support.[1]

This chapter explains how to take spatial microdata, of the type we have generated in the previous chapters, and allocate the resulting individuals into household units.

As with all spatial microsimulation work, the appropriate method for household creation depends on the data available. Data availability scenarios, in descending order of detail, include:

- Access to a sample of households for which you have information about each member.
- Access to separate datasets about individuals and households, stored in independent data tables that are not linked by household ID.

---

[1] More precisely, we can cite Dominique Dubourg (OWS), Véronique Tellier (OWS), Luc Dal (DEMO), Mélanie Bourguignon (DEMO) and Jean-Paul Sanderson (DEMO).

- No access to aggregate data relating to households, but access to some individual level variables related to the household of which they belong (e.g. number of people living with, type of household).

This chapter explains methods for household level data generation in the latter two cases. The first possibility, having a sample of households, is the topic of next chapter (Chapter 11) on the TRESIS method. In this chapter, we focus on the two cases where you have no microdata for the households (meaning data with one row per household).

The chapter is structured as followed:

-*Independent data (individuals and households)* (Section 10.1) considers the case in which data on households and individuals remain completely separate.
- *With additional household's data* (Section 10.2.2) presents a strategy when having additional data to the individual data.

Note that the first section explains a method of the literature only theoretically, whereas the second section is developed more in detail and present results for Belgium.

10.1   Independent data (individuals and households)

When the individual level data are independent from the household level data, they can rarely be linked. Data coming from different sources, sometimes implying different total populations, can cause this inconsistency. This section describes the method proposed by Johan Barthélemy[2] for dealing with such situations.

The method is to proceed in three steps. First, we determine the individual distribution Indl, for example by using the package mipfp, as explained before. Second, we determine the distribution of characteristics for the household's data, hereafter named Hh. This can be done using the same technique as for the individual level data, considering the households instead of the individuals in the previous chapters.

Third, after individual and household level distributions have been estimated, the individuals can be allocated to households. This is done one household at a time by first selecting its type before randomly drawing its constituent members (Barthelemy and Toint, 2013).

---

[2]This is a contributed chapter by Johan Barthélemy, SMART Infrastructure Facility, University of Wollongong.

10.1.1 Household type selection

The household type selection is performed to ensure the distribution of the generated synthetic households is statistically similar to the previously estimated one, i.e. Hh. This is achieved by choosing the type $hh*$ such that the distribution Hh' of the already generated households (including the household being built) minimizes the χ^2 distance between Hh' and Hh i.e:

$$d_{\chi^2} = \sum_i \frac{(hh'_i - hh_i)^2}{hh_i^2}$$

where hh_i and hh'_i, respectively, denote the number of households of type i in the estimated and generated synthetic population. Note that this optimization is simple as the number of household types is limited.

10.1.2 Constituent members selection

Now that a household type has been determined, we can detail the members selection process. First a household head is drawn from the pool of individuals `IndPool` defined by the estimated individuals distribution `Ind`. Then, depending on the household types, a partner, children and additional adults are also drawn if necessary. This process is illustrated in Figure 10.1.

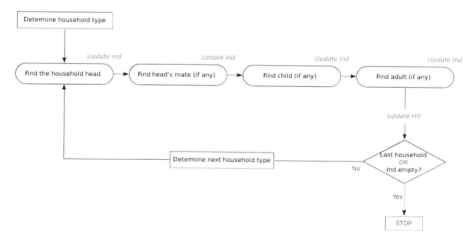

FIGURE 10.1
Constituent members selection process

Some attributes of the members can be directly obtained from their household type (for instance the gender of the head for an household of the type `Isolated Man`). The remaining missing attributes are then:

- either randomly drawn according to some known distributions (e.g. the household type x head's gender x head's age x mate's age);
- or, if different values are feasible and equally likely, retained from the distribution which minimizes χ^2 between generated and estimated distributions. This is similar to what is done for the household type selection.

After an individual type has been determined, then the corresponding member is added to the household being generated:

- if the selected class is still populated in the `IndPool`, we extract an individual from this class and add it to the household;
- else we find a suitable member by searching in the members of the households already generated. This last individual is then replaced thanks to an appropriate one drawn in `IndPool`.

Note if some additional data is available for instance the age difference between partners in a couple, then we can use it to constraint the selection of the current individual type.

10.1.3   End of the household generation process

The household generation process ends after any one of these three conditions: if all households have been constructed; if the pool of individual is empty; or if the process fails to find a member for a household in the previously generated ones. When the procedure stops, two types of inconsistencies may remain in the synthetic population: the final number of households may be smaller than estimated and/or the number of individuals estimated may be less than the known population of the area. In this case, we can form households with the remaining individuals even if they are not probable and then try to make exchanges to improve the fit. These exchanges can be made by following the principles of a `tabu search`. This consists of an algorithm that remembers the last tries to avoid repeating the same exchange many times (Hongbin Zhang, 2002).

10.2   Cross data: individual and household level information

In some cases, information about households is included in the individual dataset. For example, individual level data may include variables on type of household or/and the number of cohabitants in addition to gender and age. This provides cross-tabulated information between the households and the individuals. Considering the microdataset, IPF can help to obtain, per zone, inhabitants described by individual level variables (such as sex, age and income) and some household level information (such as household type and household's size).

To form the households with this resulting data, we have two possible alternatives. The first is to aggregate the information concerning the individuals and the households independently. By this way, we build two independent tables and we can use an algorithm similar to the one in Section 10.1. The second possibility aims to preserve the full potential of the data. This means that individuals are joined with the constraints to follow as well as possible their characteristics. For example, two people being head cannot live together; if a person has 3 cohabitants, he needs to be in a household of 4 individuals. The former solution is simpler and requires only the first chapters of the book. However it results in a loss of possible precision. The second possibility, which preserves all the information in individual and household level tables, is explained in this section.

With cross data, we usually proceed in two stages. First, we create the individuals with all their characteristics. The second step is to group these individuals into households using combinatorial optimisation. Each person must be matched to one and only one household.

For this process there are two possible methods. One assumes access to household level variables only in the individual level data. The other assumes access to additional data concerning the structure of the households such as the age difference amongst a married couple. These options are described below.

Note that in both situations, the aim is to form households where each individual is contained in one and only one household. Moreover, each individual must respect, as well as possible, its household's attributes.

10.2.1 Without additional household's data

When household level constraints are only contained in the individual's characteristics, they are often several possible groupings.

Consider the case of our Belgium study where the individual level variables are age, gender, municipality, education level, professional status, size and type of households and link with the household head (e.g. wife, child). A good grouping is one that maximises the number of well-fitted constraints. The perfect grouping would be one in which each individual respects its household size and its type of household, as well as his link with the head. In general, it is impossible to reach a perfect grouping, since the data are not perfect. Indeed, it can happen, for example, that there are an odd number of people who need to live in couple, making it impossible to find a perfect coupling.

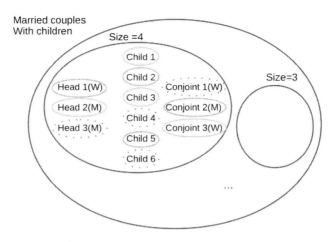

FIGURE 10.2
Illustration of the problem of grouping members of married couples with children.

As illustrated in Figure 10.2, the individuals can be categorised first by type of household ("married couples with children"" in this case) and then by size of household. This household type has a size of at least 3 (two parents and at least a child). Inside this restricted set of households, the next step is to look at the link that each individual has with the household head and again split the pool of individuals, per link. It is only after this classification that we proceed to the random draw, respecting the links.

For example, for the married couples, we first draw randomly a head and then a partner of the opposite gender (the national register of Belgium for 2011 doesn't contain homosexual couples). Then, depending on the size of household to be generated, the right number of children are also drawn. This process ends when no additional household can be drawn and respect the constraints. Figure 10.2 shows that we have a household with head 1, who is a woman; partner 2, who is a man; and two children (with ids 2 and 5).

The main sources of error with this method are incoherence in the data and error caused by the IPF process before the grouping. The method implicitly assumes that each household is equally likely to occur, independent of its characteristics.

10.2.2   With additional household's data

Without additional data on household structure, the only possible method is
the one described in Section 10.2.1. However, this allows improbable households,
such as a couple formed by an individual of 18 years old and another of 81
years. For this reason, when we create households, it is often very useful to
take into account the differences between age distributions (when these data
are available). We can consider the ages within a couple, but also of parents
and children.

To do this, we need tables of age differences. These tables are pertinent only
when considering variables already included in the simulation (for example, it
is impossible to consider a table of age differences per hair color if this variable
is not in the model). To explain the process, we develop here the methodology
used for the creation of the couples. This means that we have men and women
of different ages and roles in the household (head or spouse) and that we need
to form the couples. The random draw executed when having no additional
data will be improved by considering the real age distributions. Imagine that
a part of the additional data is the one in Table 10.1.

Municipality	Woman's age	Man's age	Count
TestCity	20-25	15-20	4
TestCity	20-25	20-25	25
TestCity	20-25	25-30	18
TestCity	20-25	30-35	8
TestCity	20-25	35-40	2
. . .	. . .	. . .	. . .

TABLE 10.1: Example of an age distribution table for couples
without children.

Note that this is a fictive table, non corresponding to any real data, just
to explain the reasoning. Thanks to this table, we know that to fit the real
population, we will need 25 couples with a man and a woman, both in the
same age class 20-25, etc.

However, these data being not perfect (because coming from different sources
with very little variations in the counts or because the synthetic population
is not perfect), the marginals could be incoherent with the ones from current
synthetic population. For this reason, we will consider the new information
only as proportions. For our example, it means that in the total of women
having 20-25 years old (57 individuals), $\frac{4}{57} = 0.07 = 7\%$ are married with a

man of age 15-20. With this reasoning, we can calculate the new Table 10.2, with a supplementary column considering the proportions.

Municipality	Woman's age	Man's age	Count	Proportion (%)
TestCity	20-25	15-20	4	7
TestCity	20-25	20-25	25	43.9
TestCity	20-25	25-30	18	31.6
TestCity	20-25	30-35	8	14
TestCity	20-25	35-40	2	3.5
...	...	...	...	...

TABLE 10.2: Example of an age distribution table with the proportion of men married with a woman of each age.

These proportions will be useful in next step of the global process. The methodology for the couples with male heads is illustrated in Figure 10.3. For female heads, the process is totally similar.

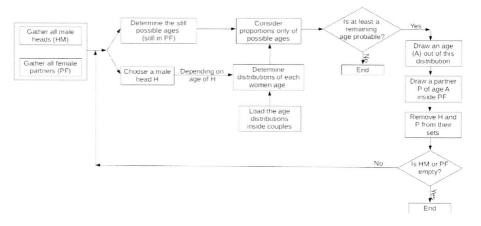

FIGURE 10.3
Illustration of the algorithm to form the couples

First, we split the set of individuals depending on their role and gender. This forms male heads and female partners to join on one hand, and female heads with male partners on the other hand. We consider each male head turn by turn. For each head, we determine the theoretical distributions of each women ages, depending on the age of the head (thanks to the additional age distribution table). Out of this distribution, we remove the ages that are no more available in the set of possible partners. Indeed, at the end of the process, only a few partners remain to be assigned. Thus, all ages will not be represented any more. Out of this distribution of ages, we calculate the proportions, whose will be used as probabilities in the random draw. Then, thanks to this, we draw an age. Finally, knowing the age of his partner, we choose a wife randomly. This process is repeated until the set of remaining individuals is empty, or there are no remaining partners in the possible ages for the remaining heads.

In our research, this algorithm has been applied to all municipalities in Belgium. The final result is illustrated in Figure 10.4. On this graph, each point corresponds to a combination of (age woman x age man) for a municipality. Its abscissa is the theoretical count for this category, included in the database of the age distributions inside couples. Its ordinate is the number of couples in this category in our synthetic population. Since the dots are on the line formed by the points having both coordinates equal, we can argue that our simulation worked well.

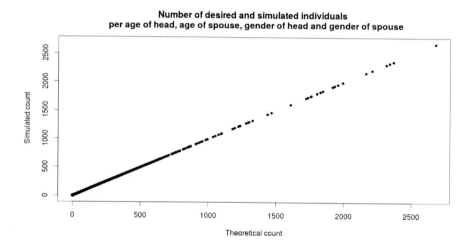

FIGURE 10.4
Illustration of the results for the couples in Belgium

The assembling of children to the head has been made by a similar process and gives as good results in terms of distribution of ages. However, when no new combination is still possible, some individuals could remain without an assigned household. To improve the spatial microsimulation here, we have chosen to join remaining individuals without regarding at their size of household if this improves the age distribution. This implies less people without an household, but some individuals are in a household not corresponding to its size.

Figure 10.5 indicates that the worst municipality has only 0.5% of non-assigned individuals. The vast majority of the Belgian municipalities has less than 0.1% of these individuals.

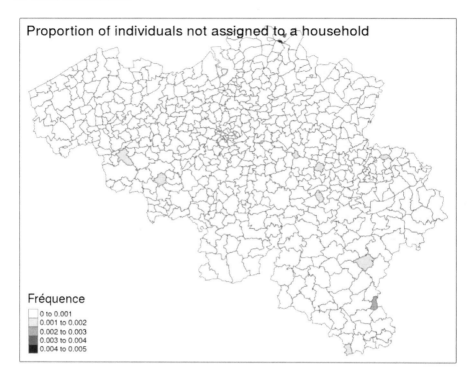

FIGURE 10.5
Illustration of the non-assigned individuals in Belgium

Figure 10.6 illustrates the proportion of individuals living with a wrong number of cohabitants. In the different municipalities, the proportion varies from 0.7 to 1.05%. These errors affect only a small proportion of each municipality and concern only the size of household (the type of household is always respected).

In conclusion, the simulation fits well the strong contraints (age distributions inside couples and between head and children). The individuals assigned to a household have the right link with the head and most of them live with the right number of cohabitants. Only few individuals has not been chosen to form an household. Thus, we can consider the synthetic population acceptable. Thanks to spatial microsimulation, these new synthetic data are statistically very similar to the real Belgian population. Since these individuals are 'synthetic', the resulting population doesn't suffer of privacy law problems.

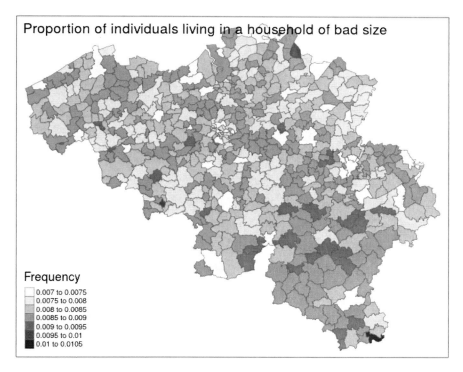

FIGURE 10.6
Illustration of the individuals in a household of a different size for Belgium

Note that a simulated annealing could be another method to resolve this kind of problems. In our case, we have tested it, but it takes very long CPU time to obtain a result as accurate as the one shown above. However, in cases where the objectives are different, it is possible that a simulated annealing becomes better. Indeed, for our purpose, it worked well, but only the computational time was a major drawback. If you would like to fit age distributions, diploma distributions, and more complicated cases, the simulated annealing could become a good option.

10.3 Chapter summary

This chapter has described methods for allocating individuals into households. Depending on the precision of the available data, the process can result in synthetic households that are more or less representative of the real population. The more pertinent information available, the more realistic will be the resulting households.

Part III

Modelling spatial microdata

11

The TRESIS approach to spatial microsimulation

CONTENTS

This chapter, by Richard Ellison and David Hensher,[1] presents an alternative approach to spatial microsimulation that involves both the generation and simulation of individuals in household units. The method, known as TRESIS, has been developed for many years at The University of Sydney. TRESIS is mature, well tested and has been designed at the Institute of Transport and Logistics Studies (ITLS) to help solve transport-related problems.

This is the first published implementation of one component of TRESIS in R. It does not use a package of its own (although the development of a **TRESIS** package implementing the TRESIS approach in R would be possible, building on the following code). Except for **dplyr**, **stringr** and **mlogit** (for data manipulation, string processing and multinomial logit models, respectively), the methods are written entirely in base R. Therefore, if you understand the

---

[1]Richard B. Ellison and David A. Hensher are based at Institute of Transport and Logistics Studies, The University of Sydney Business School, The University of Sydney.

R language well, this chapter can also be seen as a stand-alone introduction to the TRESIS method, from a technical standpoint.

The chapter is divided into four sections: an overview of the method with, reference to academic research (Section 11.1); a demonstration of the generation of realistic households used in TRESIS, building on CakeMap input data (Section 11.2); and an illustration of how demand models can be used to allocate these synthetic households to geographic zones (Section 11.3). The work concludes with a brief discussion of the merits of the approach and directions that one could develop the methods highlighted.

11.1   Overview of TRESIS modelling system

TRESIS is a Transport and Environment Strategy Impact Simulator originally developed by Professor David Hensher and others at the Institute of Transport and Logistics Studies (ITLS). TRESIS evolved from an earlier project for the Bureau of Transport and Communications Economics (Hensher, 2002; Hensher and Ton, 2002). The TRESIS software provides users with the ability to assess the impacts of a large number of transport and land-use policies on various systems linked to travel behaviour. These include the environment, user expenditure and government revenue. TRESIS incorporates an extensive model system including discrete and continuous choice models. These are combined with a process of population (or more accurately, household) synthesis and a set of equilibration criteria. TRESIS provides a powerful alternative to some more widely used microsimulation implementations for evaluating strategic projects (Hidas, 2005).

Building on the practical focus of the book, the bulk of the substantive material focusses on the generation of synthetic households, which is described in detail. The methods and code blocks outlined below are not provided as 'production ready' methods but stepping stones towards further transport modelling work in R. Notwithstanding recent efforts such as the **stplanr** package,[2] R is currently little-used in transport research, despite its power and flexibility. The chapter demonstrates how the synthetic populations generated through TRESIS can be used to identify where different types of households live using a residential location model. The chapter concludes with discussion of other important aspects of transport modelling that could be integrated into R, building on this work.

Overall, the TRESIS model system was designed to cover many of the long and short-term decisions made by individuals and households that affect how

---

[2]See `https://github.com/ropensci/stplanr` for more information.

they travel. These range from the very long-term (e.g., where to live and where to work) to the very short-term (e.g., departure time and travel mode). When used together, the TRESIS models predict households' long-term decisions and how these decisions then affect how individuals within that household travel given user-specified transport and land-use policies. Although in this chapter we discuss how to implement a stand-alone residential location model, the full power of the TRESIS system is only available when the whole set of demand models is estimated and run together. The figure below shows an overview of the high level TRESIS model structure (Hensher, 2008).

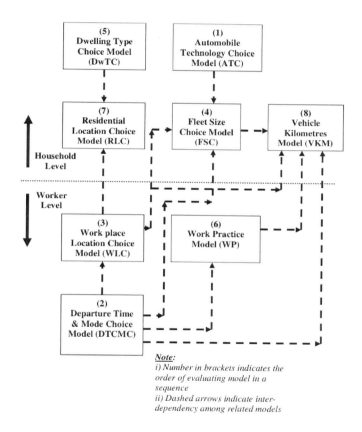

FIGURE 11.1

Overview of TRESIS model structure

The TRESIS demand models have been estimated from a combination of stated preference (SP) and revealed preference (RP) data using the NLOGIT software (see Hensher et al. (2015) for details of the software and estimation models). The model system provides for feedback between the different models using inclusive value parameters. These inclusive value parameters provide the mechanism through which the decisions made using one model (e.g., what vehicle to buy) are incorporated into the decisions made using other models in the system (e.g., where to live). In this chapter, we discuss only the stand-alone residential location model. This is just one part of TRESIS, which would be closely coupled to other sub-models covering fleet size, vehicle type choice, workplace location and dwelling type choice, among others (Hensher, 2008).

11.1.1   Differences between TRESIS and other microsimulation systems

TRESIS differs from standard micro-simulation models in that it was designed primarily as a strategic model and in that it preserves the behavioural richness of observed individuals. In contrast, standard micro-simulation techniques commonly used in transport are largely designed for small-scale studies that cover only a small area and are typically focused on traffic simulation rather than behavioural simulation (Hidas, 2005). TRESIS tends to operate at higher geographical levels than micro-simulation systems. Individuals in the household are not run through a micro-simulation model to determine specific routes. Instead, the TRESIS system estimates travel demand based on separate travel demand models within a network assignment model. Although this means that link level travel times for a specific individual cannot be determined, it provides more reliable aggregate estimates of travel time across the whole network. These are needed for the efficient running of the TRESIS models.

11.2   Synthetic households

11.2.1   What are synthetic households?

Synthetic households as used in TRESIS are a combination of a set of socio-demographic characteristics and a weight that represents their incidence in the population. These synthetic households form the building blocks for TRESIS and provide the necessary underlying variables required for the models in the simulation. The households are sampled such that given certain combinations of socio-demographic variables, there is a sufficiently large variation in the sample to reasonably cover all segments of the population that are of interest.

Synthetic households in TRESIS

There are fundamental differences between the populations generated through traditional population synthesis (described in earlier chapters) and synthetic households in TRESIS. First is that synthetic households retain their original composition and characteristics from the source sample. Rather than assigning sample individuals to households, in TRESIS individuals remain linked to their original households. Households are seen as a single entity, and used as the unit of analysis for many of the models, including the residential location model, within TRESIS. This has the side effect that each household is not simply a sum of its members but may also be considered to have its own characteristics.

Second is that synthetic households in TRESIS are not assigned to a unique geographic zone. Instead, the sample from which the synthetic households are drawn is limited to those households that are resident in the whole of the study area (generally a city's Greater Metropolitan Area). This results in synthetic households having an overall geographic coverage matching the study area but no specific location within it. This approach means that the weights of the synthetic households are calculated based on their incidence in the entire population in the study area and not just their specific geographic location. The specific location of synthetic households is then assigned by running them through the TRESIS model and equilibrating the model system at trip, location and vehicle type levels. This reproduces with some accuracy the base year spatial travel and location profiles, as well as vehicle type and fleet size composition. Among others, TRESIS also includes a model for the choice of residential location using more disaggregate zones. It must be emphasised that household weights do not change when they are assigned to zones.

11.2.2 Required data for generating synthetic households

Like IPF and other population synthesis methods discussed previously, the generation of synthetic households in TRESIS requires a combination of micro level census data and aggregate population statistics.

The microdata needs to include matched household and individual level data. These are retained within the synthetic households. The aggregate population data should include the variables that are of most relevance to the research question for selecting synthetic households. The only *requirement* from the aggregate level data is that it includes a total count for the whole study area. The census data should ideally be available as a multi-way cross-tabulation for each combination of variables that are desired for representing the households in the target population. The reasons for this are explained further later in this chapter. For now suffice to say that the input data for TRESIS are all available from the census or other reliable sources in most advanced nations such as Australia, where the model was first developed for six of the state capital cities. The input data could come from different sources, however, if the variables are comparable.

11.2.3   Synthetic households in R

Generating synthetic households in R, using the TRESIS method, requires three stages. The first involves ensuring that the variables in the micro-data sample are compatible with those used in the demand models. Where variables used in the demand models or in the aggregate population level data are derived from one (or more) of the original micro-data variables, then these variables must be defined for each of the households in the micro-data sample. The second stage involves selecting variables on which to base the synthetic households and then calculating their weights and required sample. Third, the micro-data is sampled to generate the final synthetic households each with an individual weight based on their incidence in the population.

For this chapter, we will use the same CakeMap dataset as was used in the previous chapter. To ensure that we are using the base data without any modifications, re-import the CakeMap data files:

```
ind <- read.csv("data/CakeMap/ind.csv")
cons <- read.csv("data/CakeMap/cons.csv")
```

Recall that the CakeMap dataset records individuals and not households. This means we need to add a variable to the dataset that records which household each individual is in. Since we do not know who in the CakeMap dataset is in which household, for this example we will randomly assign each individual to a household:

```
set.seed(99) # This line sets the seed for the random number
# generation in the sample() function used below. This is
# used here to ensure that the results are consistent for each
# run. To ensure that your weighted of synthetic variables is
# reasonable for your application it is best to run this
# several times without a seed and compare the results.

# Randomly select the sizes of the households and then assign
# the unique IDs of each household to sequential individuals
# in the dataset.
ind$hhld <- rep(1:1000, times=sample(c(1:6), 1000, replace=TRUE)
               )[1:nrow(ind)]

# Check how many households we have in our dataset:
length(unique(ind$hhld))
```

```
## [1] 260
```

Keep in mind that the sample() function randomly selects numbers so you will likely have a different number of households in your dataset unless you set the seed using *set.seed(99)* as above. Once we have assigned each individual to a dataset we can generate the household level data. For this example we can simply calculate some of these household variables including the number of individuals in the household, the number of workers and the age of the oldest person living in the household. Ensure you have loaded the **dplyr** and **stringr** packages and then run:

```
library(dplyr)
library(stringr)


hhlds <- group_by(ind, hhld) %>%
  summarise(
    numres = n(),
    # Values of 8 and 97 mean "long-term unemployed" and
    # "not answered" respectively
    numworkers = length(NSSEC8[NSSEC8 < 8]),
    nummanager = length(NSSEC8[NSSEC8 == 1.1]),
    numprof = length(NSSEC8[NSSEC8 == 1.2]),
    numrout = length(NSSEC8[NSSEC8 == 7]),
    # Take the midpoint of the age ranges then take the maximum
    # for that household
    maxage = max(
      rowMeans(
        read.table(
          text=str_replace(ageband4, '-', ','),
            header=FALSE, sep=',')
      )
    ),
    # Value of 1 means they have a car, 2 means they do not.
    # Divide the number of cars by 1.5 to get a more realistic
    # number since more than 1 person can use the same car.
    numcars = floor(sum((Car-2)*-1)/1.5)
  )

hhlds[,c("hhld","numres","numworkers","nummanager",
        "numprof","numrout","numcars")]


## Source: local data frame [260 x 7]
##
##      hhld numres numworkers nummanager numprof numrout numcars
```

##	(int)	(int)	(int)	(int)	(int)	(int)	(dbl)
## 1	1	4	4	0	0	1	1
## 2	2	1	1	0	0	0	0
## 3	3	5	5	0	0	0	2
## 4	4	6	5	0	0	1	3
## 5	5	4	4	0	0	2	2
## 6	6	6	5	0	0	2	2
## 7	7	5	4	1	0	1	2
## 8	8	2	2	0	0	1	1
## 9	9	3	3	0	0	1	2
## 10	10	2	2	0	0	0	0
## ..	...	...	...	...	...	...	...

This code uses **dplyr**'s group_by() and summarise() functions to generate the household level data from the individual level data. Generally, some household level data is already provided in the micro-data sample from the census but it may be necessary to add additional summary variables using the individual level micro-data.

The *cons* data frame from the CakeMap dataset contains the aggregate population level data we need for generating the weights and identifying the required sample. However, since the *cons* data frame is at the Ward level, we need to aggregate to get a total for all the wards. This is easy to accomplish by using the colSums() function:

```
totcons <- colSums(cons)
```

Now that we have finished preparing the data, we can start generating the synthetic households. Just as with selecting the constraint variables in the previous chapter, we first need to choose which variables are the most important to ensure that our synthetic households reflect the distribution of these variables in the population. We should also select the number of households we want to generate. Which variables should be used depends on exactly what is being investigated. If it is of interest to identify how a specific policy will affect households with a different number of workers (e.g., peak hour congestion charging), then it is important to ensure that the number of workers is used as one of the variables in the generation of synthetic households. For this relatively simple dataset, we will use the availability of a car and the number of professionals in the household as the weighting and sampling variables. Typically, synthetic households in TRESIS are generated using three (nested) pairs of variables including household income, number of workers, family structure (e.g., family with children, family without children, single-person household, etc) and the occupation of the reference person, among others. However, since the cross-tabs are not available for the CakeMap data, for the purposes of this example we will assume that the distributions of the two variables (car and professionals) are the same for all combinations as they are for the population. Similarly, we will assume that the distribution for households is the same for individuals.[3]

First, let us look at the distributions of the two variables across the population using the *prop.table()* function:

```
# Car vs No Car:
prop.table(c(totcons['Car'],totcons['NoCar']))
```

```
##    Car NoCar
## 0.701 0.299
```

```
# Professionals vs All others:
prop.table(c(Professionals=totcons['X1.2'],
          OtherOcc=sum(
            totcons[c(match('X1.1', names(totcons)),
                     match('X2', names(totcons)):match(
                       'Other', names(totcons)
                     )
                   )])))
```

```
## Professionals.X1.2        OtherOcc
##             0.0595          0.9405
```

The *prop.table()* function takes two or more values and generates a table with the proportion of each value is of the total. The *match()* function is used to find the elements in the vector, *names(totcons)* containing the names of the variables in the totcons data frame, that match the relevant variable name. This is used here to avoid hard coding the correct index which we may change inadvertently. Using this code we see that approximately 30% of households in the population have no car and that only 6% of the population are professionals (occupation category 1.2). Given our assumption that these distributions also apply for all combinations, we can calculate that for the population as a whole using the *tcrossprod()* function:

```
popdistprop <- tcrossprod(
  prop.table(c(totcons['Car'],totcons['NoCar'])),
  prop.table(c(Professionals=totcons['X1.2'],
            OtherOcc=sum(
              totcons[c(
```

---

[3]This will generally not be the case in reality and so every effort should be made to source multi-way cross tab data for the required variables.

```
                    match('X1.1', names(totcons)),
                    match('X2', names(totcons)):match(
                      'Other', names(totcons)
                      )
                  )])))
)
colnames(popdistprop) <- c('Professionals','Other')
rownames(popdistprop) <- c('Car','NoCar')
popdistprop
```

```
##         Professionals Other
## Car            0.0417 0.659
## NoCar          0.0178 0.281
```

This code uses the *tcrossprod()* function to take the two pairs of proportions we used earlier and multiplies all values in the first matrix by each value in the second. In this case this results in a 2x2 matrix that shows the distribution for the combination of the two variables. Keep in mind that this would not need to be a square matrix if the variables used have more than two values. From this we can calculate the actual number of households in each combination. Since the CakeMap dataset includes only the number of people rather than the number of households, we will need to make the assumption that each household has a certain number of people. If we assume that each household has an average of three people living there then we would use the code below to calculate the total number of households in each group in the population as a whole. Note that if you have the actual number of households in each group, this should be used instead.

```
popdisttot <- round(
    popdistprop*(totcons['Car']+totcons['NoCar'])/3
  )
popdisttot
```

```
##         Professionals  Other
## Car             22577 356923
## NoCar            9624 152143
```

If we choose to generate 100 synthetic households, we simply need to multiply the matrix by 100 and then round to see how many households we need to sample from each of the for combinations of variables:

```
popdist <- round(popdistprop*100)
popdist
```

```
##         Professionals Other
## Car               4    66
## NoCar             2    28
```

In this case, each group has at least one household in the sample. However, when more variables are used (or variables with more values), it is common for several groups to be too small to have a sufficient proportion in the population to be represented in the synthetic households (particularly with only 100 households). When this occurs, these small groups are combined into a single larger group such that together they are represented by at least one household.

The last step before we sample the synthetic households is to generate the weights. Similar to other other methods of population synthesis discussed earlier in this book, the weights tell us how many real households each of our synthetic households represent in the population. This is calculated by simply dividing the total number of households in each group by the number in our set of synthetic households:

```
popweights <- popdisttot/popdist
```

You may choose to round the weights to ensure that each synthetic household represents only whole households but this is not absolutely necessary.

We can now sample the households with which to generate the synthetic households. Although it is possible to automate this process provided the variable names are mapped to each other, we will do this manually to illustrate the process. Taking the first group (professionals with a car) we can select the required number of households by using the following code:

```
synhhlds <- sample(
  filter(
    hhlds,numprof > 0, numcars > 0)$hhld,
  popdist['Car','Professionals']
  )
synhhldweights <- rep(
  popweights['Car','Professionals'],
  popdist['Car','Professionals']
  )
synhhldtypes <- rep(
  'Car+Professionals',
  popdist['Car','Professionals']
  )
```

The second line takes the calculated weights for that group of synthetic households and repeats it for the required number of households using the *rep()* function. The third line is used just to help identify from which group each household was sampled. This is generally not used but it can be useful for checking the results. Once we have sampled all the households we will bind the two vectors (*synhhlds* and *synhhldweights*) into a single data frame. Note that the use of the sample() function means that the specific households chosen will likely differ somewhat every time unless the *set.seed()* function is used first (as we did above). This is why it is important to select the correct variables because the only variables that are guaranteed to match the distribution in the population are those that are used to select the synthetic households. This is then repeated for each of the subsequent groups.[4]

This results in a vector with the household IDs of the households from the micro-data selected to be used for the synthetic households and a separate vector for the household weights. Since these two vectors are both in the same order, we can create a data frame that contains the synthetic household ID and the matching household weights.

```
synhhldDF <- data.frame(hhld=synhhlds,
                        hhldweight=synhhldweights,
                        hhldtype=synhhldtypes
                        )
```

Once this has been done generating the synthetic household dataset is relatively straightforward requiring us to merge the individual and household level data for our desired households. In TRESIS itself, this is followed by some additional work to ensure that any variables not required for the models are removed as well as making any necessary changes to ensure confidentiality of the micro-data is retained.

```
synhhlddata <- merge(
  filter(hhlds, hhld %in% synhhlds),
  filter(ind, hhld %in% synhhlds),
  by='hhld',
  all.x=TRUE,
  all.y=TRUE
)

# Merge with the weights data frame
synhhlddata <- merge(
  synhhlddata,
```

---

[4]The code for this stage is over 50 lines long so is not included in the book. See the online version (https://github.com/Robinlovelace/spatial-microsim-book/blob/master/11-Tresis_chapter.Rmd) of the chapter of http://git.io/vGAEz for the code.

```
  synhhldDF,
  by='hhld',
  all.x=TRUE
)


# Add an ID for each individual in the household
# (starting from 1 for each household)
synhhlddata <- group_by(synhhlddata, hhld) %>%
  mutate(
    indid = 1:n()
  )
```

The *synhhlddata* data frame now contains all the data we need to use within the residential location model for assigning households to zones. So we can avoid regenerating synthetic households every time we change our residential location model, it is useful to save our data frame to an RData file:

```
save(synhhlddata,file="output/synhhlddata.RData")
```

11.3 Using demand models to allocate synthetic households to zones using R

The key to the TRESIS system are its demand models. As discussed earlier in this chapter, the models used in TRESIS include a variety of discrete choice models covering many of the travel-related decisions made by individuals and households. The models used in TRESIS are linked meaning the model for residential location cannot be completely separated out from the rest of the models. However, the TRESIS approach can be used even with a single stand-alone residential location choice model and given its (potential) simplicity, it is this approach that will be discussed here.

At its most basic, a residential location model as used with the TRESIS synthetic households is a discrete choice model that predicts, up to a probability, in which area each household will choose to live given their unique set of socio-demographic characteristics and the characteristics of the different zones. Although the full details of the estimation of discrete choice models is outside the scope of this book, it is worth briefly discussing how the models used for TRESIS can be estimated. The discrete choice models used for residential location in TRESIS have been estimated from disaggregate choice data collected using a combination of revealed preference and stated preference surveys. It is also possible to estimate discrete choice models using micro level data provided sufficient spatial detail is available for your purposes. The models in TRESIS were estimated using the NLOGIT software package because it provides features needed for advanced models unavailable in most other packages. However, many discrete choice models as would be used by those developing residential location models can also be estimated in R using the **mlogit** and **mnlogit** packages.

11.3.1   Simple discrete choice model for residential location

The CakeMap data we are using does not have the disaggregate data we need to estimate the discrete choice model. For this reason, we will use a "model" where the parameters have been randomly selected. This means that our results in this case are unlikely to reflect the true values provided in the *cons* data frame. Nonetheless, the procedure we will use is the same as would be used if a properly estimated discrete choice model was available.

The discrete choice model that is needed is one where each alternative (zones in our study area) have an estimated utility function. The utility functions for each alternative can include either the same independent variables or different independent variables as well as common parameters or alternative-specific parameters. Since our dataset contains 124 wards, we will simplify our example by reducing these to 4 larger zones of 31 wards each. Zone 1 will include wards 1-31, zone 2 will include wards 32-62, etc. The model below was estimated based on using average values for each ward using a very simple model and cannot be regarded as a "good" model. However, it fits the required structure for the utility expressions and in the absence of disaggregate data for estimating the model we will use it for this example.

$$U_1 =$$
$$3.3 * mgmt + 1.85 * rout + 1.73 * numCars$$
$$U_2 =$$
$$6.56 - 1.35 * prof + 26.81 * mgmt - 0.83 * rout - 3.11 * numCars$$
$$U_3 =$$
$$0.12 - 1.08 * prof + 1.4 * mgmt + 1.78 * rout + 1.99 * numCars$$
$$U_4 =$$
$$3.71 - 8.09 * prof + 33.24 * mgmt + 1.81 * rout - 1.29 * numCars$$

In this model, the subscript on U indicates the zone, *mgmt* is the number of workers in the household who are managers, *prof* is the number of workers who are professionals, *rout* is the number of workers whose occupation is considered "route", and *numCars* is the number of cars in the household.

Applying the model requires the values of each of the synthetic households to be run through the utility equations to calculate the probabilities of the households living in each of the four zones. The easiest way to calculate the utilities for each synthetic household is to make use of R's vector functionality. Since our model includes only household level characteristics, we begin by simplifying the synthetic household data to one row per household with only the variables necessary for our model, making sure to rename them to match the variable names in our utility expressions:

```
synrun <- unique(synhhlddata[,c("hhld","nummanager","numprof",
                                "numcars","numrout",
                                "hhldweight","hhldtype")])
synrun <- rename(synrun,
                mgmt = nummanager,
                prof = numprof,
                numCars = numcars,
                rout = numrout
                )
```

We then create variables for each of the four utility expressions. The while loop iterates through each of the utility expressions to calculate the relevant utilities.[5] The *eval()* and *parse()* functions are used together to take text input and evaluate the expressions as if R code had been typed into the console or a standard R script file. The *with()* function allows the variable names in the utility expressions to be used as-is, without pre-pending the data frame name and a $ to them and not attaching the data frame to the global environment.

```
synrun <- synrun %>% mutate(
    u1 = NA,
    u2 = NA,
    u3 = NA,
    u4 = NA
)
i <- 1;
while (i <= 4) {
    synrun[,7+i] <- with(synrun,
          eval(parse(text=c(
            "3.2989*mgmt + 1.8519*rout +
            1.733*numCars",
            "6.5633 - 1.3535*prof + 26.8106*mgmt - 0.8307*rout -
            3.1094*numCars",
```

---

[5]It is possible to extract the coefficients and variables names directly from the output of mlogit if desired. The text format is shown here since models estimated using nlogit are generally written in this format.

```
    "0.115 - 1.0771*prof + 1.3984*mgmt + 1.781*rout +
    1.9871*numCars",
    "3.7052 - 8.0874*prof + 33.2434*mgmt + 1.8073*rout -
    1.291*numCars"
    )[i])))
  i <- i + 1
}
```

We can then calculate the probabilities of each synthetic household living in each zone. The equation to estimate the probabilities depend on the exact model specification that is used in the model. In this case, where the model includes alternative-specific variables (and values), we can use the equations below. These probabilities can then be used to help identify where households sharing the characteristics of each synthetic household are likely to live:

```
synrun <- synrun %>% mutate(
  p1 = exp(u1)/(exp(u1)+exp(u2)+exp(u3)+exp(u4)),
  p2 = exp(u2)/(exp(u1)+exp(u2)+exp(u3)+exp(u4)),
  p3 = exp(u3)/(exp(u1)+exp(u2)+exp(u3)+exp(u4)),
  p4 = exp(u4)/(exp(u1)+exp(u2)+exp(u3)+exp(u4))
)
```

The final probabilities for 10 households are shown in the table below. In this very simple example, you will see that there are some households with the same probabilities. Given a fully developed residential location model as the one used in TRESIS, with more explanatory variables as well as a wider range of grouping variables for the synthetic households, the number of households with the same probabilities would be very low. However, even from this relatively simplistic example, it is apparent that even within the same groups, the probabilities of living in a particular zone vary (sometimes substantially).

hhld	hhldweight	hhldtype	p1	p2	p3	p4
1	5408	Car+Other	0.22	0.08	0.29	0.41
4	5408	Car+Other	0.31	0.00	0.69	0.00
5	5408	Car+Other	0.37	0.00	0.60	0.03
6	5408	Car+Other	0.37	0.00	0.60	0.03
7	5408	Car+Other	0.00	0.00	0.00	1.00
10	5434	NoCar+Other	0.00	0.94	0.00	0.05
13	5408	Car+Other	0.10	0.56	0.14	0.20
16	5408	Car+Other	0.22	0.08	0.29	0.41

hhld	hhldweight	hhldtype	p1	p2	p3	p4
18	5434	NoCar+Other	0.00	0.94	0.00	0.05
20	5408	Car+Other	0.22	0.08	0.29	0.41

11.3.2   Results

The final step is to assign the number of households in each zone using the probabilities and the household weights. Although in some applications the highest probability is assumed to be the chosen zone, in the TRESIS approach, the probabilities are used in combination with the weights to identify how many real households with the same characteristics as the synthetic households would live in each zone.

```
synresults <- synrun %>%
  group_by(hhldtype) %>%
  summarise(
    zone1 = round(sum(p1*hhldweight)),
    zone2 = round(sum(p2*hhldweight)),
    zone3 = round(sum(p3*hhldweight)),
    zone4 = round(sum(p4*hhldweight))
)
```

The results of this (very basic) model are shown in the table below. These initial results do not accurately reflect the shares in the population in each zone. This is for a number of reasons. The primary reason is that the example model we used here was by many measures a poor model with very low goodness of fit, including a very low R^2 value and a highly insignificant χ^2 p-value. If a better model had been estimated using disaggregate data, the results would be much closer to the actual population shares.

hhldtype	zone1	zone2	zone3	zone4
Car+Other	77737	38230	137451	103505
Car+Professionals	12075	3172	7327	3
NoCar+Other	524	130409	560	20651
NoCar+Professionals	52	9551	20	1

The second reason for the difference is that the model we used was not calibrated on the population shares. The TRESIS models are calibrated to ensure they reproduce the shares in the base period before being used to run any analysis. Calibration is not a straightforward process but involves adjusting the alternative-specific constants in the model until the estimated shares match the population shares within an acceptable level of tolerance.

Although we will not run the calibration procedure here, to illustrate the sensitivity of the results to the quality of the model and the need for calibration, we will re-run the calculations with small changes to the alternative-specific constants of the utility expressions. If we decrease the value of the alternative-specific constant of zone 2 so that the utility expressions become:

$$U_1 =$$
$$3.299 * mgmt + 1.852 * rout + 1.73 * numCars$$
$$U_2 =$$
$$5.56 - 1.35 * prof + 26.8 * mgmt - 0.830 * rout - 3.11 * numCars$$
$$U_3 =$$
$$0.115 - 1.08 * prof + 1.40 * mgmt + 1.78 * rout + 1.99 * numCars$$
$$U_4 =$$
$$3.705 - 8.09 * prof + 33.2 * mgmt + 1.81 * rout - 1.29 * numCars$$

hhldtype	zone1	zone2	zone3	zone4
Car+Other	81765	20811	143302	111045
Car+Professionals	13107	1633	7833	4
NoCar+Other	949	112580	1022	37592
NoCar+Professionals	140	9429	53	2

Decreasing the constant for zone 2 further results in more changes. It is important to understand that the probabilities do not change linearly in response to an equal change in the value of the alternative-specific constant. Furthermore, the switch between zones is not equal across the groups. As a result, calibration involves the incremental changes of more than one alternative-specific constant until the shares are reproduced.

$U_1 =$
$$3.30 * mgmt + 1.85 * rout + 1.73 * numCars$$
$U_2 =$
$$4.56 - 1.35 * prof + 26.8 * mgmt - 0.830 * rout - 3.11 * numCars$$
$U_3 =$
$$0.115 - 1.078 * prof + 1.40 * mgmt + 1.78 * rout + 1.99 * numCars$$
$U_4 =$
$$3.71 - 8.09 * prof + 33.2 * mgmt + 1.81 * rout - 1.29 * numCars$$

hhldtype	zone1	zone2	zone3	zone4
Car+Other	84377	9412	147087	116048
Car+Professionals	13726	710	8137	4
NoCar+Other	1549	87167	1685	61743
NoCar+Professionals	368	9111	141	5

11.4 Conclusions

This chapter has described the process required for generating and applying the synthetic households used within the TRESIS model system to a more general application of population synthesis. Although TRESIS is a land-use and transport simulator, the approach described here is a generalised one that is applicable to a wide range of model systems that require, or would benefit from, the use of disaggregate data for simulating decision making involving spatial decisions including residential location but also workplace location and other activity destinations.

11.4.1 Limitations

The primary limitation of the TRESIS approach to population synthesis is its heavy reliance on detailed disaggregate household and population level data. This means that acquiring the necessary data needed for generating the synthetic households and estimating and calibrating the models is frequently a time-consuming and sometimes expensive undertaking. In many contexts such data simply does not exist.

However, once the data has been collected, the resulting model system is very powerful, allowing predictions not only of the current state, but also of future forecasts of decisions under uncertainty. Provided the variables on which the forecasts rely are available in the synthetic households and models and are trustworthy, the TRESIS approach provides an integrated framework for modelling transport decisions.

Freight models

TRESIS has until recently been focused entirely on passenger transport with freight transport considered as part of the capacity of the network but not considered in the demand models. As such, the methods developed for TRESIS have been optimised for passenger transport modes. However, there is ongoing development within the Institute of Transport and Logistics Studies to develop freight and light commercial vehicle models. These models, although different in some ways to the passenger models, have been found fit in well with the concept of synthetic households. In contrast to the passenger models that rely on synthetic households, the freight and commercial vehicle models have applied the concept used for generating synthetic households, to firms resulting in the development of "synthetic firms". Just as synthetic households are a collection of socio-demographic variables, synthetic firms are a collection of firm level (and worker) characteristics. The data requirements of synthetic households are however also problematic for synthetic firms.

11.4.2   MetroScan-TI

TRESIS, although extremely powerful in many ways, has lacked some features that are increasingly being needed for a full assessment of potential transport and infrastructure projects. MetroScan-TI is the latest evolution of the TRESIS model system that is currently being developed and incorporates all the behavioural richness of the original TRESIS models and extends this to include a much more detailed zone structure, a fully integrated network assignment model, freight models, models for firm location and modules for cost-benefit analysis and wider economy impact assessments. MetroScan-TI is intended to provide users with the ability to quickly and easily prioritise a large number of potential projects on a wide range of variables. These projects include the standard transport infrastructure (new railway lines, roads and service improvements) as well as other infrastructure and strategic level property development (the mix and number of flats, semi-detached and detached houses).

11.4.3 Extending residential location to transport models in R

Residential location models are an important component of any fully integrated transport model because they provide the "productions" (i.e., demand) for travel. However, at a minimum transport models also require a method of estimating the likely destinations (or "attractions") of trips, as well as the mode used for these trips. The TRESIS approach uses the same synthetic households, and the individuals within those households, as were used in this chapter for the residential location model, to estimate the destinations and modes of trips. This means the development of a full transport model in R can apply the synthetic household method described in this chapter. Although developing the full set of models cannot be described as a simple process, as demonstrated in this chapter, the functionality of R, and some of its packages, makes it well suited to applying transport models. In particular, implementing full network assignment in R is likely to be particularly challenging, but its data structures and rapidly improving GIS functionality, some of which has been demonstrated in this chapter, means these are also likely to be possible.

11.5 Chapter summary

This chapter has provided an introduction to the TRESIS approach to population synthesis through the use of synthetic households and demand models. In it we learned to generate, allocate and model household level data. TRESIS is an approach that is relatively simply conceptually but that incorporates powerful behavioural features. This, combined with its ability to model a wide range of phenomena and behaviours in the synthetic households, makes TRESIS well suited to implementation in R (subject to data requirements). It can also be applied to a variety of other applications not limited to those involving transport models.

12

Spatial microsimulation for agent-based models

CONTENTS

This chapter, contributed by Maja Založnik, is the most advanced chapter of the book.[1] It covers one of the most enticing and potentially useful extensions of the methods developed in the book: agent-based modelling (ABM). Agent-based models consist of interacting agents (frequently - but not necessarily - representing people) and an environment that they inhabit. The chapter uses NetLogo, a language designed for ABM. This will be introduced in due course, but prior knowledge of it would be advantageous.

Agent-based modelling is a powerful and flexible approach for simulating complex systems. Agent-based models (ABMs) allow analysis of problems that are highly non-linear and *emergent*, in which the dominant processes guiding change can only be seen after the event. This allows ABMs to tackle a very wide

---

[1]Maja is based at the Oxford Institute of Population Ageing, University of Oxford.

range of problems: "agent-based modelling can find new, better solutions to many problems important to our environment, health, and economy" (Grimm and Railsback, 2011).

ABMs can broadly be divided into two general categories. Discussions in the literature are more nuanced, but a broad categorisation is adequate here. First are descriptive models. These explore a system at an abstract level, in an attempt to better understand the underlying dynamics that drive the system. Second are predictive models. These attempt to represent the system to a certain level of accuracy in order to predict or explore future states of the system. One of the main barriers to the success of predictive models is reliable, high-resolution data. ABMs require information about the individual people, households or other units on which the population of agents can be based. Using spatial microsimulation to generate realistic input populations is one way that predictive ABMs can be made to more closely represent the population under study.

Although ABMs can be very diverse, all agent-based models involve at least three things (Castle and Crooks, 2006):

1. a number of discrete agents with characteristics
2. who exist in an environment
3. and who interact with each other and the environment.

Using the synthetic spatial microdata created in previous chapters we can represent the individuals and their characteristics, and allocate them to zones (the environment). With the right tools spatial microdata can be used as an input into ABM. If your aim is to use spatial microdata as an input into agent based models, you may already be more than half way there!

12.1   ABM software

A wide range of software is available for ABM. Of these, **NetLogo** is one of the most widely used for model prototyping, development and communication (Thiele et al., 2012). Alternative open source options include **Repast** and **MASON**. Both are written in Java, although Repast also includes optional graphical development and some higher-level language support. Due to its excellent documentation, ease of learning and integration with R, NetLogo is the tool of choice in this book and should be sufficiently powerful for many applications.

NetLogo (`https://ccl.northwestern.edu/netlogo/`) is a mature and widely used tool-kit for agent-based models. It is written in Java, but uses a derivation of the Logo language to develop models. The recently published **RNetLogo** package provides an interface between R and NetLogo, allowing for model runs to be set up and run directly from within R (Theile 2014). This allows the outputs of agent based models to be loaded directly into your R environment. Using R to run a separate programme may seem overly complicated for very simple models. For setting up and testing your model we recommend using NetLogo, with its intuitive graphical interface. However, there are many advantages of using NetLogo from within R, the foremost being R's analysis capabilities: "It is much more time consuming and complicated to analyse ABMs than to formulate and implement them" (Theile et al. 2012).

ABMs commonly represent unpredictable behaviour through probabilistic agent rules. This means that two model runs are unlikely to lead to identical results. Therefore we generally want to study many model runs before drawing conclusions about how the overall system operates, let alone real world implications. Much of the time taken for agent based modelling is consumed on this sensitivity/parameter space analysis. Although NetLogo has some inbuilt tools (e.g. the Behaviour Space) that support the running of large numbers of experiments, the capabilities for subsequent analysis are limited. R provides a much more comprehensive suite of data analysis tools, making the integration of R and NetLogo an obvious choice.

12.2 Setting up SimpleWorld in NetLogo

The NetLogo programming language is a well-developed and thoroughly documented modelling environment. We advise you familiarise yourself with the language, e.g. through an exploration of the numerous tutorials (`http://ccl.northwestern.edu/netlogo/docs/tutorial1.html`) and the rich library of pre-written models that come with the programme (File > Model Library) before undertaking the practical elements of this chapter. This chapter is not meant as a NetLogo primer. But following the step-by-step instructions should result in a fully functioning AMB, that can be controlled and alaysed using R. The following sections (12.2 to 12.4) introduce the basics of NetLogo's functionality to create an ABM of SimpleWorld. **RNetLogo** is introduced in Section 12.5.

In NetLogo the design of the model's graphical user interface (GUI) is handled separately from the model procedures. This is analogous to the separation in R's **shiny** (`http://shiny.rstudio.com/`) package between the user interface and the 'server side'. The latter are constructed in code using the NetLogo programming language in the Code tab (discussed below). The graphical interface, which includes the environment and the control *widgets*, is set up using the drop-down menus located in the Interface toolbar.

The whole model is saved as a .nlogo file. This text file includes code, GUI parameters and the contents of the (optional) Info tab. Although .nlogo files are human readable, most of the information relating to the GUI is difficult to manipulate. We advise against modifying the .nlogo file directly.

12.2.1   Graphical User Interface in NetLogo

The GUI of your model in NetLogo has three types of elements: controls, settings and views. The minimum requirement for a model to work — and these are the only two default elements when you open a new model — are the Graphics window (a *view*) and the Command Centre (a *control*). For most purposes however you will want to add others and adjust the defaults. The Graphics window for example, which defines the ABM *world*, has a default size of 33 by 33 patches, 13 pixels square. These wrap both horizontally and vertically, settings that can quickly turn out to be inappropriate.

The main elements of the NetLogo GUI are described below, along with their setup for the SimpleWorld agent-based model.

Views

As already mentioned, the Graphical window is the most important — mandatory — part of the GUI. We will adjust it to suit SimpleWorld with the Settings button on the right-hand side of the Interface tab toolbar (Figure 12.1). First change the default location of origin from "Center" to "Corner". Then set the value of max-pxcor to 35 and max-pycor to 17. With [0 0] describing the coordinates of the bottom left patch, this means we now have a 36 by 18 size world. Untick both the world wrapping boxes and click Apply. You may find the size inappropriate for your screen in which case adjust the value of Patch size until you are happy with the interface size. Right-clicking on a patch and selecting the bottom option "inspect patch x y" will open a patch monitor where you can inspect the values of all the patch's variables, currently only the default ones: its coordinates, colour, label and label colour. These can be changed in the monitor, but we will change them programmatically below.

In addition to the main view of our agents' world, the other optional views are monitors, plots and an output window.

Monitors report the values of variables. Our model currently has no variables. To demonstrate, add a monitor — either by right-clicking somewhere on the empty interface or by clicking the Add button on the toolbar and selecting

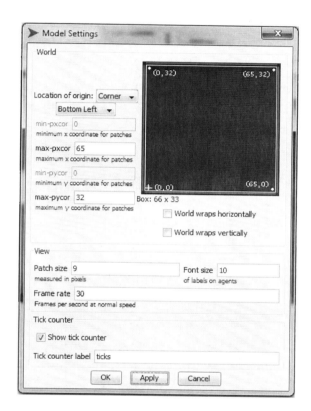

FIGURE 12.1
Setting the world size and resolution

Monitor from the dropdown menu next to it. In the dialogue box that appears type `count patches` in the reporter field and click OK.

Instead of just reporting the current value of a variable, plots report their changing value through time when the model is running, which can be a very useful visual summary of the model's progress (Figure 12.2). We will set up a plot later on, when we have populated the model with some agents! Finally the output box can be used to output text while the model is running.

Controls

The command centre can be used to issue commands on the fly, either to a finished model (even while it is running) or when developing one or testing. We can try it out to begin to design SimpleWorld by typing in the following commands:

```
observer> ask patches [set pcolor 2]
observer> ask patches [if pxcor < 24 [set pcolor 4]]
observer> ask patches [if pxcor < 12 [set pcolor 6]]
observer> ask patch 6 9 [ set plabel "Zone 1" ]
observer> ask patch 18 9 [ set plabel "Zone 2" ]
observer> ask patch 30 9 [ set plabel "Zone 3" ]
```

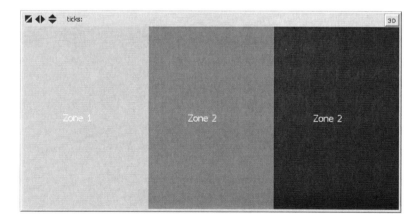

FIGURE 12.2
SimpleWorld in NetLogo

You will probably find the font size a bit small, in which case you can adjust it in the Graphic Window Settings menu. When we now save, close and reopen our model we will however find that our world — although of the right dimensions — is again black without any zone labels. The Command centre is useful for testing out code snippets like this, but we will have to include these lines in the code tab for them to be saved and run every time.

The second main type of control are the buttons. Although buttons cannot do anything you could not do in the Command centre it is usually convenient to create buttons to call your most used commands and procedures. Most models will therefore have at least a "Setup" button (which is normally run only once at the beginning) and a "Go" button, which continuously executes the commands until it is de-pressed. The commands that are run by the buttons can be simple commands such as the patch colour ones we used above, but more than likely they will be more complex procedures we will define in the code tab. It is good practice therefore to try to keep model procedures in the code tab, so that can be found one place. Alternatively code snippets can go directly inside the button dialogue boxes or run them directly from the Command centre.

Let's create a button calling the command "setup". NetLogo will offer a warning "Nothing named setup has been defined." Now switch to the Code tab and type in the following:

```
to setup
  clear-all
  create-zones
  reset-ticks
end

to create-zones
  ask patches [set pcolor 2 ]
  ask patches [if pxcor < 24 [set pcolor 4 ]]
  ask patches [if pxcor < 12 [set pcolor 6 ]]
  ask patch 6 9 [ set plabel "Zone 1" ]
  ask patch 18 9 [ set plabel "Zone 2" ]
  ask patch 30 9 [ set plabel "Zone 3" ]
end
```

The definition of the **setup** procedure, like all procedures, requires a **to** opening call and an **end** to close. All the commands within these two lines will be executed whenever we click on the setup button. The first and last lines are standard in any setup procedure as it will normally be called after a previous model run, so we want to make sure the world is back in starting position and the time or tick counter is set back to zero. The third line calls the **create-zones** procedure, which is defined underneath. We could have put all the zone-creation commands directly into the setup procedure. But in order to keep the code nicely organized and easier to read, we follow this structure, and separate out logical blocks of code. We can now test the setup button. Note that this procedure is now saved as part of the model.

Settings

There are four types of settings widgets: sliders, choosers, switches and input text boxes. These can be used by the user to change specific variables' values. By adding one of the types of settings widgets we define the *global variable* that the slider or other type of input modifies, and these variables can then be called in the code. We do not really need to define any selectable inputs at the moment, but let us assume we might want to populate the agents with attributes from different files, and create a *chooser* that can be used to select the .csv data file. In the dialogue that opens when we add the chooser we define the global variable `csv` and give it (for now only) a single possible value "agents.csv" (see Figure 12.3).

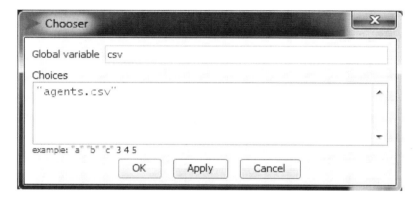

FIGURE 12.3
SimpleWorld in NetLogo

12.3   Allocating attributes to agents

In order to assign the relevant attributes to the agents we first need to create an appropriately formatted file to be read by NetLogo. The list of 33 agents and their attributes is in the data frame `ints_df`. In order to read it in NetLogo we first need to export it as a .csv file (making sure the row and column names are not also exported) and place it in the same folder as our .nlogo model:

```
write.table(ints_df, "NetLogo/agents.csv", row.names=FALSE,
            col.names=FALSE, sep=",")
```

12.3.1 Defining variables

In NetLogo we can define three types of variables:

- *Global* variables are at the highest level and are accessible to all agents.
- *Agent* variables have a unique value for each agent.
- *Local* variables are defined and accessed only within procedures.

We have already defined the global `csv` variable when we created the chooser, the other option is to declare a new global variable in the code using `globals [ 'variable-name']`. Some agent variables are predefined, for example the patches are a type of agent and we have already encountered the variables `pcolor` and `plabel`. A user defined patch variable is declared using the command `patches-own`. Let's fix the `create-zones` procedure so that in addition to colouring the patches with correct zones, we also add a zone variable to each patch and give it the correct value:

```
patches-own [zone]
to create-zones
  ask patches [set pcolor 2 set zone 3]
  ask patches [if pxcor < 24 [set pcolor 4 set zone 2]]
  ask patches [if pxcor < 12 [set pcolor 6 set zone 1]]
  ask patch 6 9 [ set plabel "Zone 1" ]
  ask patch 18 9 [ set plabel "Zone 2" ]
  ask patch 30 9 [ set plabel "Zone 3" ]
end
```

If we now inspect an individual patch, we will see that it has a new variable `zone` that holds the correct value for each patch.

Similarly we can define the variables for the agents. The generic agents in NetLogo are called `turtles` and hence the `turtles-own` command is used to add agent variables. We can also create our own *breed* of turtles and call them for example `inhabitants`. The definition of the new breed has two inputs: the name of the *agentset*, that is the set off all the agents in the breed and the name for a single agent of that breed. Once we define the breed we can declare its variables as well:[2]

---

[2]Although this is not strictly necessary, one recommended structure for your code tab is to break it up into three blocks: 1. Declarations, where you define the global variables, the breeds if any and the agents-own variables; 2. Setup procedures where you define `setup` and all its sub-procedures and in a similar vein; 3. Go procedures.

```
breed [inhabitants inhabitant]
inhabitants-own [id zone.original age sex income history]
```

These are the variables from the .csv file. Note that we changed the name of the zone variable because we already have a zone variable that belongs to the patches and, for clarity, because the agents will be able to move around SimpleWorld and it might be useful at some later stage to explore their starting zone. We have also added a variable called `history`, to record each individual agent's history for future analysis.

Local variables are defined and accessible only within procedures. For clarity — although this is not necessary — local variable names will commence with %. Reading the code we can then immediately see if a variable is of locally limited scope or if it is a global or agent variable. To keep the code even clearer, we can make sure that we always declare the global, patches-own and other agent variables at the beginning of the code, before we start writing the procedures. Assignments are then made to them using the `set` command. Local variables are declared on-the-fly (while the model is running) using the `let` command, and once declared assignments are also made to them using the `set` command.

12.3.2 Reading agent data - Option 1

Now we need to create a procedure to read the agent data. The `read-agent-data` procedure needs to:

1. open the file, read a single line of data-
2. create an `inhabitant`
3. correctly assign the five values to the five inhabitants-own variables
4. and place the agent into the correct zone.

We will also assign the agent a colour based on their gender. Then the procedure must repeat this until all the lines of data are read. The first option for accomplishing this is a simple construction, relying on the fact that our current .csv file has a fixed width file format. We know the exact position of each value, because each variable has a constant number of characters.

The NetLogo procedure therefore begins and ends with the `file-open` and `file-close` commands, thus opening a connection to the file in question and allowing us to read the data from it. We also use this opportunity to set the default shape of the agents as *person*. The main loop is a `while` construction, reading through each individual line and saving it into a local variable (only available within the procedure) `%line`.[3] We then create a single inhabitant

---

[3]Using a percentage sign in front of the names of local variables is a purely stylistic decision and makes the code easier to read. It makes it immediately obvious what the scope

and assign the substrings from `%line`, based on the position of each value, to the appropriate inhabitant's variables. We also initialize the agent's history as an empty list: [] and at the end we set the agent's colour based on their sex and position them on an unoccupied patch in the correct zone.

```
to read-agent-data
  set-default-shape inhabitants "person"
  file-open csv
  while [not file-at-end?] [
    let %line file-read-line
    create-inhabitants 1 [
      set id read-from-string substring %line 0 1
      set zone.original read-from-string substring %line 2 3
      set age read-from-string substring  %line 4 6
      set sex substring %line 8 9
      set income read-from-string substring %line 11 15
      ifelse sex = "m" [set color yellow] [set color green]
      move-to one-of patches with [not any? turtles-here and
        zone = [zone.original] of myself ]
    ]
  ]
  file-close
end
```

Making sure that we add the **read-agent-data** procedure to the setup procedure, we can now test our code simply by clicking the setup button on the graphic interface tab. We still have the monitor there counting the patches; let's edit it to change the reporter instead to something more informative: **count inhabitants** should do it. Hopefully there are 33 (see Figure 12.4)!

12.3.3   Reading agent data - Option 2

The method of reading in the data described above would not work if SimpleWorld inhabitants had, for example, a five-figure income. A more generic way of reading the data would therefore be to extract the substrings based on the positions of the commas rather than their absolute position. This method is similar to that described above, but it has an extra loop within. Having extracted a single `%line`, we first add an extra comma to the end of it (using the command **word**), which will allow us to determine the end position of the last value. The inner loop then finds the position of the first comma and saves

---

of each variable is. In NetLogo local variables are created on-the-fly using the **let** command, while agent and global variables need to be declared explicitly. The values of all variables are changed using the **set** command.

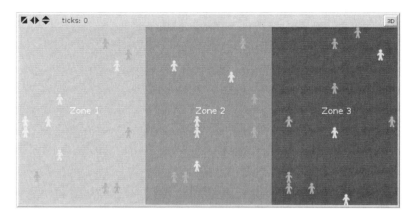

FIGURE 12.4
SimpleWorld populated with 33 inhabitants

it in %pos, reads the value between positions 0 and %pos and saves it as %item, and appends %item to an internal list variable called %data.list. It then removes this item from %line, shortening it by one element, and repeats the loop. This while loop runs until all the items have been extracted individually and saved in %data.list. The rest of the code is similar to the first version: each item is assigned to the inhabitant's variables, their colour is determined and finally they are positioned in their correct zone.

```
to read-agent-data-2
  set-default-shape inhabitants "person"
  file-open csv
  while [not file-at-end?] [
    let %case file-read-line
    set %case word %case ","
    let %data.list []
    create-inhabitants 1 [
      while [not empty? %case] [
        let %pos position "," %case
        let %item read-from-string substring %case 0 %pos
        set %data.list lput %item %data.list
        set %case substring %case (%pos + 1) length %case
      ]
      set id item 0 %data.list
      set zone.original item 1 %data.list
      set age item 2 %data.list
      set sex item 3 %data.list
      set income item 4 %data.list
      ifelse sex = "m" [set color yellow] [set color green]
```

```
      move-to one-of patches with [not any? turtles-here and
         zone = [zone.original] of myself ]
      ]
   ]
   file-close
end
```

To test this option, we simply add `read-agent-data-2` to the `setup` procedure. Make sure you comment out the other read agent procedure - otherwise NetLogo will simply run both of them, populating SimpleWorld with two sets of our 33 inhabitants.

12.4   Running SimpleWorld

We will now create a simple set of rules for the inhabitants of SimpleWorld. At each *time tick* the inhabitants will:

1. move to a random location within their zone.
2. "look across the fence": check their field of vision for inhabitants from a neighbouring zone and select the closest one in view.
3. try to "convince" them to come over to the other side: the inhabitant with more money (`income`) will *bribe* the other with 10% of their money to come over to their zone.

The model will have the following adjustable parameters:

1. The field of vision has two parameters: the viewing angle and the distance
2. Average level of *bribeability* of inhabitants: if their level is less than 100%, a random number generator will be used to determine whether the agent accepts the bribe or not. The distribution of bribeability is approximately normal with a mean and a standard deviation.

12.4.1   More variable definitions

We first need to define these four global variables using settings widgets. To do this use the add button to add the following four sliders to the model:

global variable name	minimum	increment	maximum	value	meaning
angle-of-vision	0	10	360	100	
distance-of-vision	0	1	10	4	
average-bribeability	0	1	100	100	
stdev-bribeability	0	1	20	0	

The graphic interface area with the aforementioned widget settings should now look approximately like Figure 12.5.

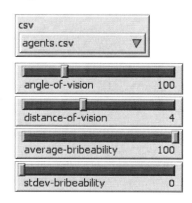

FIGURE 12.5
New global variables defined using sliders

The average bribeability and its standard deviation that we have just defined will translate into an individual value for each inhabitant. We need to define an agent variable for that purpose. Furthermore, during the bribe negotiations, we need each inhabitant to be in communication with a maximum of one other inhabitant. In order to make sure that happens, we will define a boolean variable to keep track of that. We must therefore add these two variables to the `inhabitants-own` line (now written over several lines only for clarity):

```
inhabitants-own [
  id
  zone.original
  age
  sex
  income
  history
  bribeability
  in-negotiations]
```

12.4.2 More setup procedures

Before we start constructing the Go procedure, we need to add one more setup procedure: one that assigns each agent a random level of bribeability. We will sample from a normal distribution with the mean and standard deviation defined using the sliders. One way of doing that is with the following line:[4]

```
ask inhabitants [set bribeability random-normal
  average-bribeability stdev-bribeability ]
```

The problem with this command, is that the normal distribution is unbounded. But bribeability values below 0% or above 100% make no practical sense. We therefore use the following little trick to force the values into the correct interval:

```
to determine-bribeability
  ask inhabitants [
    set bribeability median (list 0 (random-normal
      average-bribeability stdev-bribeability) 100) ]
end
```

Placing the random value in a list with the values 0 and 100 means we can use the **median** command to extract it when it falls within the interval. But if it falls below 0 the median value of the three will be 0, and similarly if the random normal is over 100, the median will be 100. We can now add the procedure **determine-bribeability** to the setup procedure, which should be as follows:[5]

```
to setup
  clear-all
  create-zones
  ;read-agent-data
  read-agent-data-2
  determine-bribeability
  reset-ticks
end
```

Try it out with different values for the average and standard deviation of bribeability and make sure it never goes outside the bounds of meaningfulness.

---

[4]You can try it out simply by typing it into the command centre. Seemingly nothing will happen, but if you inspect an inhabitant you will see that their **bribeability** variable is now 100.

[5]Make sure you keep clear-all as the first and reset-ticks as the last procedures.

12.4.3   The main Go procedure

We are now ready to write the main action procedure, normally labelled the **go** procedure. To do that we create a button in the same manner as before, this time running the command **go**, and make sure we tick the **Forever** tickbox in the dialogue box. This means clicking the 'go' button (with a circular arrow on the face) will execute the command *repeatedly*, until we stop it manually or until an end condition is met. Let's first set up the bare bones of the go procedure: in each time step we want the inhabitants to **reposition** themselves, to **engage** in the bribery negotiations and we want to **record** their states in their **history** variables. Finally we want the tick counter to advance by a single time step:

```
to go
  ;reposition
  ;engage
  ;record
  tick
end
```

Because we haven't defined the first three procedures yet, we are keeping them commented out, as soon as we define them, you should delete the semicolon and test the code.

For repositioning the inhabitants we can use a very similar line of code used to first position the inhabitants in SimpleWorld. To reposition all agents we can use the **ask** command. NetLogo implements all **ask** commands sequentially, going through all agents or patches in *random order*.

```
to reposition
  ask inhabitants [
    move-to one-of patches with [not any? turtles-here and
      zone = [zone] of myself ]
  ]
end
```

If we now uncomment **reposition** in the go procedure, we can test it out. You can adjust the speed of the command execution with the slider on the toolbar at the top of the window.

The construction of the **engage** procedure is a bit more complex, so let's break it down into steps:

1. **reset**: first we must reset everyone's **in-negotiations** variable to **false**

2. **find-pairs-and-negotiate**:

- 2.1 find all viable negotiating pairs: ask each inhabitant (not already in negotiations) if there are any viable candidates in their field of vision that are in a different zone. And if so, define these two inhabitants as `party` and `counter-party` respectively.

- 2.2 then such pairs must `negotiate` i.e. compare incomes, determine the winner, transfer the funds and reposition the losing inhabitant into the winner's zone.

So we therefore need (keeping the unwritten procedures commented out for now):

```
to engage
  ;reset
  ;find-pairs-and-negotiate
end
```

The `negotiate` procedure will be called from `find-pairs-and-negotiate` and is the last part we will need. For now we will create a place-holder, so we can test the code as we go along. A negotiation will always take place between two inhabitants (labelled the `party`) and the `counter-party`and `parties` for the turtle-set containing both. It will first of all turn their `in-negotiation` switch to true, so we know they cannot engage in any other negotiations. We will expand the code to include the actual *bribed migration* further down. We will only turn their colour red, so we can see who (if anyone) is in negotiations.

```
to negotiate [ parties ]
  ask parties [
    set in-negotiations   true
    set color red]
end
```

Next, let's write the reset procedure, which is very straightforward: at the beginning of each new time step, everyone's `in-negotiation` switch is turned back off and everyone's colour is changed back to what it was before:

```
to reset
  ask inhabitants [
    set in-negotiations false
    ifelse sex = "m" [set color yellow] [set color green]]
end
```

The final procedure is `find-pairs-and-negotiate` which relies on several local variables (prefixed by the % symbol). The pseudo code is as follows:

1. for each inhabitant that is not already in negotiations

- define %field-of-vision as the patches in their sight and also in a neighbouring zone
- define %viable-partners as all the inhabitants on %field-of-vision, who are available for negotiations

  2. if there is at least one inhabitant in %viable-partners, we have a match!

- define the active inhabitant as %party and the closest one of the the possible viable partners as %counter-party
- now that we have both parties to the negotiations, we can run the negotiate procedure on them

The NetLogo code is as follows:

```
to find-pairs-and-negotiate
  ask inhabitants [
    if in-negotiations = false [
      let %field-of-vision patches in-cone distance-of-vision
         angle-of-vision  with [zone != [zone] of myself]
      let %viable-partners ((inhabitants-on %field-of-vision)
         with [ in-negotiations = false])
      if count %viable-partners > 0 [
         let %party self
         let %counter-party  min-one-of %viable-partners
            [distance myself]
         negotiate (turtle-set %party %counter-party)
         ]
    ]
  ]
end
```

You should now be able to test the code by uncommenting (by removing the semicolon ; symbol) the engage and go procedures. To actually observe what is happening at each time step, you can edit the go button and untick the Forever option. This way each time you press go, you will get a single repositioning and a single run of engage. You can now already start using two of the sliders: changing the angle and distance of vision of the inhabitants will change the number of inhabitants that engage in negotiations on average. We still have a little count inhabitants monitor in the graphics window, which is not very informative. Edit it instead to change the code to:

```
count inhabitants with [ in-negotiations = true ]
```

as well as changing the display name to something shorter. If you've done everything right the monitor should only ever show even numbers!

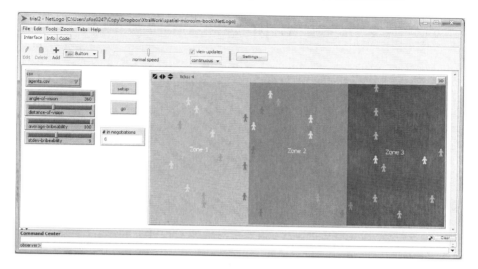

FIGURE 12.6
SimpleWorld in negotiations

Figure 12.6 is a screenshot of a state of SimpleWorld with 8 inhabitants engaged in active negotiations. Only they are not actually negotiating yet. That is the procedure we left unfinished before and need to rewrite now that we have everything else working. We will break it down into pseudo-code again, in each negotiation we :

1. first establish who is the winner and who is the loser using `sort-on` on their incomes (if there is a tie, `sort-on` selects a random winner)
2. using the loser's `bribeability` level and a random number generator determine if they will accept the bribe
3. if so, then transfer 10% of the winner's income to the loser, and transport the loser into the winner's zone.

Let's return to the `negotiate` procedure we created above as a place-holder and add the actual negotiation to it. We will leave the first three lines from before setting both parties `in-negotiations` switch to true and colouring them red. The next three lines sort them by income and assign them to either `loser` or `winner`. This is followed by an `if` construct, that only executes if the loser's bribeability is larger than a random number between 0 and 100. In this case the loser takes the bribe, is moved into the winner's zone and their colour changes to brown. At the same time the bribe must also be subtracted from the winner's income.

```
to negotiate [ parties ]
  ask parties [
    set in-negotiations   true
    set color red]
  let loser-winner sort-on [income] parties
  let loser item 0 loser-winner
  let winner item 1 loser-winner
  if (random-float 100 < [bribeability] of loser)[
    ask loser[
      set income income + 0.1 * [income] of winner
      move-to one-of patches with [not any? turtles-here and
        zone = [zone] of winner]
      set color brown]
    ask winner[
      set income income * 0.9]
    ]
end
```

We should now have a working model. Go ahead and try it out, e.g. by loading SimpleWorldNetlogo.netlogo.nlogo (`https://github.com/Robinlovelace/spatial-microsim-book/blob/master/NetLogo/SimpleWorldVersion2.nlogo`). After clicking the setup button, you will probably want to adjust the speed in the toolbar to make the model run a bit slower, then simply press Go!

Note that we still have an undefined command `record` in the go procedure. Depending on what type of analysis we anticipate, we might want to have a full history of what happened in each model run to each of our inhabitants. We can therefore make sure that at every tick the inhabitant's `history` variable (a list we initialized in the `read-agent-data` procedure) gets appended with the agent's income and zone. We can use the command `lput` to append the current values to the list:

```
to record
  ask inhabitants [
    set history lput (list who ticks income zone ) history]
end
```

Add this procedure to the code and make sure you uncomment `record` in the go procedure. We will return to the history variable in the final section of this chapter.

12.4.4 Adding plots to the model

After a few repeated model runs the little icons moving around the zones and changing colours will probably become less interesting. You may want another way of visualising what is happening. To do this we can create a couple of plots to summarise what is going on. We will add a line plot that keeps track of the average income in each zone, and another plot displaying the number of inhabitants in each zone. Using the add button from the toolbar we can add a plot somewhere in the empty space in the graphical interface, while making sure there is enough room for both the plots we are planning. A dialogue box will open, where we can first rename the plot to *Average Income*. Ticking the show legend box is optional. Change the axis ranges to 0 - 500 for the x axis and 2000 - 3000 for the y (don't worry, as long as you have the Auto-scale box ticked, the plot will automatically adjust when the lines fall out of range). In the *pen update commands* add the code to plot the mean income of zone 1 inhabitants, which first makes sure there are actually inhabitants in the zone:

```
if any? inhabitants with [zone = 1] [
plot mean [income] of inhabitants with [zone = 1]]
```

It is easier to add the code if you click on the pencil icon on the right, where the dialogue box that opens makes it easier to see. Then click on the colour on the left and select a colour and finally change the *Pen name* to zone 1. Now simply click on the **Add pen** button and repeat the process for zones 2 and 3, and click OK. You will probably want to increase the size of the plot, by right-clicking it, choosing select and then manipulating the corners of the plot. Now you can re-run the model while the plot updates!

The plotting options in NetLogo are not very extensive. It is possible to create more involved plots using the plotting primitives enabling, for example, an area graph. As before let's create a new plot and name it *Population*. Change the axis ranges to 0 - 500 and 0 - 33. Now define the pens as before, but only assign them the correct colour and name. Instead of adding individual pen update commands as before, we will write the plotting commands into the box above, the **plot update commands**. The reason, as you will see, is that we want the three pens to draw successively, not concurrently (Figure 12.7). We will use the following commands:

- **set-current-plot-pen** "name" specifies which pen we are currently using - the options being the names we gave the pens in the previous step
- **plot-pen-up** and **plot-pen-down** lift or lower the pen from the plotting area - where pen is by default set down
- **plotxy** "value1" "value2" moves the pen to the position determined by the x and y coordinates - depending on whether the pen is down or up, it will either draw or not draw.

In addition we will use a local variable, %total, to keep track of the cumulative total of inhabitants as we go through the three zones, starting from zone 3:

```
let %total 0
set-current-plot-pen "zone 3"
plot-pen-up
plotxy ticks %total
set %total %total + count inhabitants with [ zone = 3 ]
plot-pen-down
plotxy ticks %total
set-current-plot-pen "zone 2"
plot-pen-up
plotxy ticks %total
set %total %total + count inhabitants with [ zone = 2 ]
plot-pen-down
plotxy ticks %total
set-current-plot-pen "zone 1"
plot-pen-up
plotxy ticks %total
set %total %total + count inhabitants with [ zone = 1 ]
plot-pen-down
plotxy ticks %total
```

12.4.5   Stopping behavior

The last thing to add, to complete the model, is some sort of stopping condition. As it stands the model will keep on running until we de-press the *go* button. There are several ways of doing this, using the command stop in the go procedure. We could for example insert as the first line into the go procedure the following code: if ticks >= 1000 [stop], which will simply stop the model after 1000 iterations. Or we can stop it when a certain criteria is met, such as when all inhabitants end up single zone, using the following code:

```
if (count inhabitants with [zone = 1] = 33) or
   (count inhabitants with [zone = 2] = 33) or
   (count inhabitants with [zone = 3] = 33) or
   (ticks >= 1000)  [stop]
```

You will have noticed by now that the model, if left to run, ends up in one of two ways: either with all inhabitants in one zone, or in an indefinite "tie" between zones 1 and 3. Adding the if ticks >= 1000 condition is one way of dealing with the tie issue. But it becomes problematic if we expect the model under some conditions to take longer to reach a single zone winning, so this is

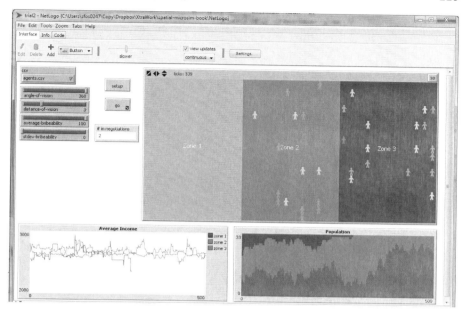

FIGURE 12.7
Plotting SimpleWorld

not ideal. Another way to avoid the problem of ties is to change the geography of SimpleWorld by *wrapping* the world horizontally (in the GUI settings menu) which creates a border between zones 1 and 3. SimpleWorld then becomes a vertical cylinder and inhabitants from zones 1 and 3 can bribe each other across their new mutual border.

But in many situations this is not a satisfactory solution and we want to preserve the geography as we originally intended it and allow a tie to be a legitimate end point of the model. So as a final addition to our SimpleWorld ABM we will add some code to keep track of the distribution of inhabitants across the zones and the number of ticks it has remained stable. Then we need to decide on a (rather arbitrary) cut off point after which we declare a tie and stop the model.

To do this we must define two global variables at the beginning of the code. The variable `inhabitant-distribution` will be a list of three values: the current number of inhabitants in each zone and the variable `time-stable` will be the number of ticks that have passed since the list has changed.

```
globals [
  inhabitant-distribution
  time-stable
  ]
```

We initialize the two variables from the setup procedure, so we add a new procedure `start-distribution-tracking` at the end of the setup procedure:

```
to start-distribution-tracking
    set time-stable 0
  set inhabitant-distribution (list
    count inhabitants with [zone = 1]
    count inhabitants with [zone = 2]
    count inhabitants with [zone = 3])
end
```

Then we have to add a `track-distribution` procedure that executes at every tick, so we add it to the go procedure - at the end, after `record`. Each time `track-distribution` is called we create a local variable `new-inhabitant-distribution` and compare it to the existing one. If it is the same, we add one tick to the `time-stable` counter, otherwise we reset the counter back to zero. Finally we replace the old distribution variable with the new one:

```
to track-distribution
  let new-inhabitant-distribution (list
    count inhabitants with [zone = 1]
    count inhabitants with [zone = 2]
    count inhabitants with [zone = 3])
  ifelse new-inhabitant-distribution = inhabitant-distribution [
    set time-stable time-stable + 1]
  [set time-stable 0]
  set inhabitant-distribution new-inhabitant-distribution
end
```

We now have a running counter tracking how long it has been since the inhabitant distribution has changed. Now all we need is to decide that e.g. 200 ticks with no change indicates a tie and the simulation can safely be stopped:

```
if (count inhabitants with [zone = 1] = 33) or
  (count inhabitants with [zone = 2] = 33) or
  (count inhabitants with [zone = 3] = 33) or
  (time-stable >= 100)  [stop]
```

Now we have a working model to explore. You can try running it with different parameters to see how they affect the behaviour of the model, and you can also try expanding the model to add more. For example a simple way of doing that is by making the 10 % bribe a parameter of the model that can be adjusted by the user. But to systematically investigate the model, we must explore the *parameter space* by carefully varying all the parameters and recording the results.

NetLogo provides a tool for this called *Behaviour Space*, which can be found in the *Tools* menu. It allows you to specify how to vary the variables, how many times to repeat each experiment and which measures to report after each run. Additionally, the tool offers the possibility of running many experiments simultaneously, taking advantage of multi-core processors. Unfortunately this way of doing a parameter sweep is not codeable — the commands defining the *experiment* are entered in a dialogue box, and the outputs are saved as .csv files, which require further handling for analysis. There are other extensions for NetLogo, such as BehaviourSearch, which can search the parameter space using various search algorithms to optimize an objective function. These and other tools may meet your requirements and/or fit with your workflow. But we will return full circle now back to R and use it to run SimpleWorld and analyse the model results using RNetLogo.

12.5   Control the ABM from R

The RNetLogo package (Thiele, 2014) allows us to control a NetLogo model directly from R. Now that we have a working model and have explored it in NetLogo to get a feel for what it's doing (and make sure that it is doing what we want it to do), we can explore the model more systematically. RNetLogo allows us to issue commands, set global values, repeatedly run the model and collect the results directly from R.

The RNetLogo package is installed in R like any other package using `install.packages("RNetLogo")`. Note that RNetLogo depends on rJava (Urbanek, 2013), which may require some extra configuration of Java, depending on your setup. The basic code for opening and running a model in R is then:

```
library(RNetLogo)
nldir <- "C:/Program Files (x86)/NetLogo 5.1.0" # program dir
# nldir <- "/home/robin/programs/netlogo-5.1.0/" # e.g. localtion
mdir <- "C:/Users/mz/NetLogo/SimpleWorldVersion2.nlogo" # model dir
# mdir <- "/book_dir/NetLogo/SimpleWorldVersion2.nlogo" # e.g.
NLStart(nldir, gui = FALSE)
NLLoadModel(mdir)

[...main code here ...]

NLQuit()
```

In the above code `NLStart` initiates an instance of NetLogo (in this case in headless mode, since the `gui` argument is set to `FALSE` — try with `gui = TRUE`) and `NLLoadModel` loads the model. Make sure you have the correct paths to the folders where you have installed NetLogo and where the model is saved. Some example directories are commented out for Linux users. The main commands we will want to run in the main code are `NLCommand()` which simply sends a command for NetLogo to execute, and `NLReport()`, which also returns a reporter (value). The commands work the same way as if we were entering them in the command center. This also means that the command `go` is only executed once i.e. not forever as it can be using the GUI button. For example, now that we have opened the model, we can run the following code from R:

```
NLCommand("setup")
NLReport("ticks")
NLCommand("go")
NLReport("ticks")
```

The first reporter should have reported the number of ticks to be 0 and the second one should 1, demonstrating that the tick counter has advanced by one. To repeatedly issue a command or report request we can use the `NLDoCommand()`, so for example:

```
NLDoCommand(50,"go")
NLReport("ticks")
```

should now show the tick counter to be 51. We can also repeatedly call a reporter, in this case we have to pass both a command to be repeated a certain number of times and a reporter that will be executed each time. Naturally we can also assign the result to an R variable, and instead of the default list data class return it as a data frame instead:

```
test <- NLDoReport(10,"go",
                    c(" ticks",
                      "count inhabitants with [zone = 1]",
                      "count inhabitants with [zone = 2]",
                      "count inhabitants with [zone = 3]"),
                    as.data.frame = TRUE)
test # show the resulting data frame

#    X1 X2 X3 X4
# 1  52 11  9 13
# 2  53 11  9 13
# 3  54 11  9 13
```

```
#  4  55 11   9 13
#  5  56 12   8 13
#  6  57 12   8 13
#  7  58 12   9 12
#  8  59 12   9 12
#  9  60 12   9 12
# 10  61 12   9 12
```

Finally we can also extract the current variable values from an agentset. For example the incomes and current zones of all inhabitants, which we can then visualize using a simple boxplot (Figure 12.8):

```
current.state <- NLGetAgentSet(c("who","income", "zone"),
                                "inhabitants")
boxplot(current.state$income ~ current.state$zone,
  xlab="Zone", ylab="Income",
  main=paste("Income distribution after",
    NLReport("ticks"), "ticks" ))
```

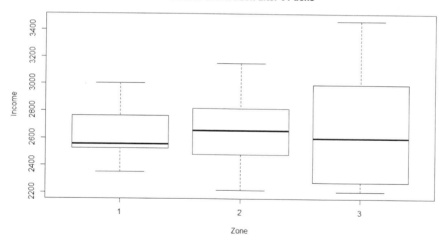

FIGURE 12.8
NetLogo data plotted in R

The `NLCommand` and `NLReport` functions can therefore be executed either singly or repeatedly using their "Do" versions. For ultimate control there are also

"Do-While" versions of the functions which take as an argument a condition and keep executing until the condition is met.[6]

To continue with our simple example the `NLDoCommandWhile` function could be used simply end the simulation after a certain number of ticks:

```
NLDoCommandWhile (" (ticks <= 100) " , "go")
NLReport("ticks")
```

12.5.1   Running a single NetLogo simulation

We now have all the ingredients to run our model from R, and probably do not need to observe the running model in the graphical interface any more. RNetLogo allows us to run the NetLogo silently in the background (in so called *headless mode*), which runs quite a bit faster than using the GUI. We can also wrap all our NetLogo commands in an R function and make further simulation more modular.

We will set up a single simulation where we open NetLogo silently, load the model, run the simulation until one zone wins or there is a definite tie, and then collect the `history` of all inhabitants at the end. There is one thing we need to keep in mind running the simulation from R though: we are repeatedly calling a *single* go procedure. This means we can still continue to call go even if the stopping conditions in the procedure have been met, it simply means the call does nothing. Because there is no warning (or error - nothing went wrong) we have no way of knowing from R that we are continuing to call go which is not doing anything - unless we check the tick counter every time and make sure it has advanced by one.

To run the simulation in R as we had designed it in NetLogo we simply use the same stopping conditions we used in the go procedure. In fact, if you plan to only run your model from R it is a good idea to delete the stopping conditions from the NetLogo code completely and make sure you always define them in the call from R. That way you can avoid setting stopping conditions in R that are less strict than the ones in the model, which would again mean R calling a go procedure that is not doing anything. And be careful: in NetLogo they were

---

[6]To prevent NetLogo from getting stuck in an endless loop if you inadvertently set a condition that is never met,both of these commands have a `max.minutes` argument, the default value of which is 10. The execution will halt after this time has passed before the condition was met. This may be relevant especially if you are expecting to run a long simulation in which case you should increase the value - of course if you are absolutely certain that the condition will be met, then you can also set `max.minutes` to 0 and the command will be executed for as long as it takes. In fact for testing your code, when you might be prone to inadvertently writing endless loops, it will make your life easier to change the value to 1 and avoid waiting the 10 minutes it takes for R to stop execution.

stopping conditions, but in `NLDoCommandWhile` they are exactly the inverse, so make sure you rewrite them appropriately!

```
SimpleWorld <- function(time.stable = 100) {
  NLCommand("setup")
  NLDoCommandWhile(paste(
    "(count inhabitants with [zone = 1] < 33) and",
    "(count inhabitants with [zone = 2] < 33) and",
    "(count inhabitants with [zone = 3] < 33) and",
    "(time-stable <= ", time.stable, ") ") , "go")
  NLGetAgentSet("history", "inhabitants")
}
```

The `SimpleWorld` function can even take an argument that gets passed on to NetLogo: here we have made the number of ticks before we declare a tie adjustable, in case we decide at some point that 100 is not really an appropriate value. Now we can run the simulation and explore the resulting data. Just like R, NetLogo also uses a pseudo random number generator so we can set the random seed we get the same result each time - this may be useful for testing your code and making sure you have set up the simulation the same way as these instructions.

```
NLQuit() # quit any existing NetLogo instances
NLStart(nldir, gui=FALSE)
NLLoadModel(mdir)
NLCommand("random-seed 42")
inhabitant.histories <- SimpleWorld(50)
```

We now have the complete history of all 33 inhabitants in the simulation to explore! We can, for example, count how many times each inhabitant moved zones during the simulation and see if that number correlates with their starting income (see Figure 12.9). You can perhaps guess inhabitants with low income will have changed more often, simply because they will have lost more negotiations, but how does the number of moves correlate with their final income? Or with their net income gain during the simulation?

```
library(dplyr)

history <- as.data.frame(matrix(unlist(inhabitant.histories),
  ncol = 4, byrow = TRUE))
colnames(history) <- c("id", "tick","income", "zone")

changes <- group_by(history, id) %>%
  mutate( change=c(0,diff(zone))) %>%
```

```
summarize(start.income = income[1],
          end.income = tail(income,1),
          income.change = end.income - start.income,
          zone.changes = sum(change != 0))

par(mfrow=c(1,3))
plot(zone.changes ~ start.income , data = changes)
abline(lm(zone.changes ~ start.income , data=changes))
plot(zone.changes ~ end.income , data = changes)
abline(lm(zone.changes ~ end.income , data=changes))
plot(zone.changes ~ income.change , data=changes)
abline(lm(zone.changes ~ income.change , data=changes))
```

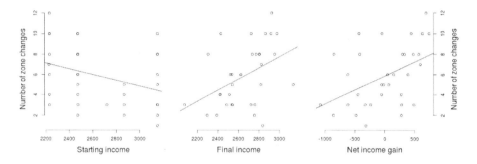

FIGURE 12.9
Number of zone changes correlated with inhabitants' income variables

Naturally there are many other things to explore in the `history` data frame.
The flip side of the collected results being so rich is of course that this is a very
large data object and its collection during the simulation and importation into
R cost valuable resources. The recommended course of action is therefore to
explore the `history` data on a single simulation to decide on what is really of
interest. Then change the NetLogo model to only collect the data we are really
interested in. In this case we could change the `record` procedure to only track
the number of zone changes for each inhabitant, which would mean simply
recovering 33 values at the end of the simulation instead of the full history
table, which in this case (a relatively short run) was over 15,000 values. When
running multiple simulations exploring the parameter space of the model, as
we will do in the next section, collecting such data would be prohibitive.

12.5.2 Running multiple NetLogo simulations

We will focus on a simple outcome variable: the amount of model time (number of ticks) it takes for the simulation to end with clear win for one zone. For simplicity we will treat all simulations that exhibit no changes for over 200 ticks as non-convergent and ignore them. We will explore how this outcome varies depending on the parameters `angle-of-vision` and `distance-of-vision`. This is called a *full factorial experiment* and is only really practical if we have a few parameters to explore, which is why we will stick with only the two variables mentioned.

First let's rewrite the simulation function to set the appropriate variables before each run (the bribeability variables will not be changing in this simulation, but it is safer to set them explicitly as they may have been changed manually in NetLogo and we would not even notice, because we are running it without the GUI!). After the variables are set, we of course call the `setup` procedure. We can omit the conditions for running `go` by simply stopping after nothing has changed for 200 ticks, regardless of how many zones are still in play. After the simulation is finished, we collect two values: the number of ticks minus the value of time-stable, giving us the actual length of the simulation run before it stabilized, and the number of zones that were occupied at the end.[7]

```
SimpleWorld <- function(angle.of.vision=360,
  distance.of.vision=10,time.stable = 200) {
  NLCommand (paste("set average-bribeability", 100))
  NLCommand (paste("set stdev-bribeability", 0))
  NLCommand (paste("set angle-of-vision", angle.of.vision))
  NLCommand (paste("set distance-of-vision",
    distance.of.vision))
  NLCommand("setup")
  NLDoCommandWhile(paste("(time-stable <= ", time.stable, ") ")
    , "go")
  c(NLReport(c("ticks - time-stable",
    nrow(unique(NLGetAgentSet( "zone", "inhabitants"))))))
}
```

The next step is to create a function to run multiple simulations where `reps` determines the number of repetitions of each parameter value combination and `a.o.v` and `d.o.v` determine the value sequences of each parameter that the simulation iterates through. Within it we first create a data frame of the parameter space we intend to examine, with a single row for each simulation run. The function then iterates through this data frame, each time running the `SimpleWorld` function with the appropriate arguments and collecting these,

---

[7]Ensure you remove the history recording from the .nlogo file as it is no longer necessary and will simply slow down the simulations.

together with the simulation results, as a data frame with a single row. At the end of all the runs, these are simply binded together into a final data frame with all the results.

```
MultipleSimulations <-
  function (reps=1, a.o.v = 360, d.o.v = c(5,10)){
  p.s <- expand.grid(rep = seq(1, reps),
    a.o.v = a.o.v, d.o.v = d.o.v)
  result.list <- lapply(as.list(1:nrow(p.s)), function(i)
    setNames(cbind(p.s[i,], SimpleWorld(p.s[i,2], p.s[i,3])),
      c("rep", "a.o.v", "d.o.v", "ticks", "zones")))
  do.call(rbind, result.list)
}
```

To be on the safe side, we may want to give this function a little test run with only two sets of parameter values and two repetitions each.

```
results.df <-
  MultipleSimulations(reps = 2, a.o.v = 360, d.o.v = c(5,10))

results.df # display the results data frame

#    rep a.o.v d.o.v ticks zones
# 1    1   360     5    49     2
# 2    2   360     5    94     1
# 3    1   360    10    16     2
# 4    2   360    10    57     1
```

The first three columns in the `results.df` data frame are simply the simulation parameters: the number of the repetition and the angle-of-vision and distance-of-vision parameters that were passed to NetLogo. The final two columns give the number of ticks the simulation required to stabilize and the number of zones occupied at the end. Assuming we are happy with the functions, we can now try to running a more serious simulation e.g. five repetitions exploring pretty much the whole of the parameter space - this may take a while!

```
results.df <- MultipleSimulations(5,seq(60,360,30),seq(1,10) )
```

With the results safely stored in the data frame, we can summarise the data and visualise it.[8] Using the `dplyr` package to reorganize our data we can see how the number of zones occupied at the end of the simulation differed across the parameter space.

---

[8]To smooth out the results the data presented here is based on 20 repetitions.

```
zones <- results.df %>%
  group_by(a.o.v, d.o.v, zones) %>%
  summarize(prop.sims = n()/reps) %>%
  group_by(a.o.v, d.o.v)

head(zones) # show the output

# Source: local data frame [6 x 4]
# Groups: a.o.v, d.o.v
#
#    a.o.v d.o.v zones prop.sims
# 1     60     1     3      1.00
# 2     60     2     2      0.15
# 3     60     2     3      0.85
# 4     60     3     1      0.15
# 5     60     3     2      0.75
# 6     60     3     3      0.10
```

We can see that in the first set of simulations with angle of vision being 60 and the distance of vision only 1, all of the simulations ended with a three-way draw. In the next set, where the distance was set to 2, 85 % of the simulations still ended in a three-way draw, while the remaining 15 % were a two-way draw. Figure 12.10 summarises the data for all 2200 simulations. While there is a clear cluster of three-way ties in the bottom left hand corner, which corresponds to the simulations with the smallest field of vision, there is interestingly no observable pattern to distinguish the two-way ties and single zone simulations.

If we focus on the simulations where a single winning zone emerged, we can find the average number of ticks required across the parameter space:

```
av.ticks <- results.df %>%
  group_by(a.o.v, d.o.v) %>%
  filter(zones == 1) %>%
  summarize(mean.ticks = mean(ticks, na.rm=TRUE))

head(av.ticks) # display the results

# Source: local data frame [6 x 3]
# Groups: a.o.v
#
#    a.o.v d.o.v mean.ticks
# 1     60     3  1876.6667
# 2     60     4  1577.1000
# 3     60     5   973.6667
```

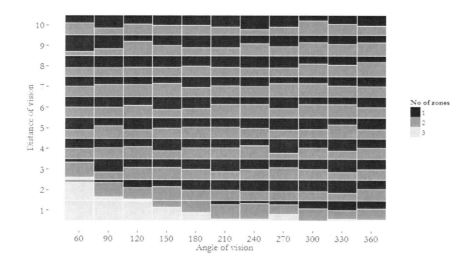

FIGURE 12.10
Number of zones occupied at end of simulation (20 repetitions)

```
# 4     60      6     368.7000
# 5     60      7     265.6000
```

The average tick counts for all the 2200 simulations (or rather in this case only the 989 that ended with a winning zone) can be plotted as a heatmap (Figure 12.11).

12.6   Chapter summary

This is chapter provides a worked example of the numerous options for agent-based modelling building on microsimulation. You have at your disposal a powerful modelling combination when combining NetLogo with R. For an excellent overview with more practical examples see Thiele et al.(2014), which covers a variety of parameter estimation and optimisation methods as well as a comprehensive set of approaches to sensitivity analysis using the RNetLogo setup described here.

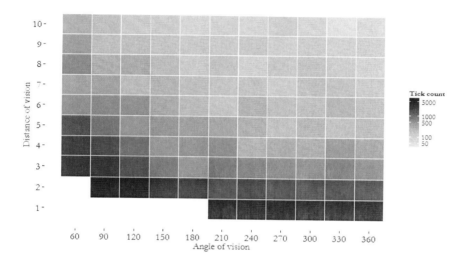

FIGURE 12.11
Average number of ticks in simulations with single zone winning

13

Appendix: Getting up-to-speed with R

CONTENTS

As mentioned in Chapter 1, R is a general purpose programming language focussed on data analysis and modelling. This small tutorial aims to teach the basics of R, from the perspective of spatial microsimulation research. It should also be useful to people with existing R skills, to re-affirm their knowledge base and see how it is applicable to spatial microsimulation.

R's design is built on the idea that everything that exists is an object and everything that happens is a function. It is a *vectorised, object orientated* and *functional* programming language (Wickham 2014). This means that R understands vector algebra, all data accessible to R resides in a number of named objects and functions must be used to modify objects. We will look at each of these in some code below.

13.1   R understands vector algebra

A vector is simply an ordered list of numbers (Beezer 2008). Imagine two vectors, each consisting of 3 elements:

$$a = (1, 2, 3); b = (9, 8, 6)$$

To say that R understands vector algebra is to say that it knows how to handle vectors in the same way a mathematician does:

$$a + b = (a_1 + b_1, a_2 + b_2, c_3 + c_3) = (10, 10, 9)$$

This may not seem remarkable, but it is. Most programming languages are not vectorised, so they would see $a + b$ differently. In Python, for example, this is the answer we get:[1]

```
a = [1,2,3]
b = [9,8,6]
print(a + b)
```

```
## [1, 2, 3, 9, 8, 6]
```

In R, the operation *just works*, intuitively:

```
a <- c(1, 2, 3)
b <- c(9, 8, 6)
a + b
```

```
## [1] 10 10  9
```

This conciseness is clearly very useful in spatial microsimulation, as numeric variables of the same length are common (e.g. the attributes of individuals in a zone) and can be acted on with a minimum of effort.

13.2   R is object orientated

In R, everything that exists is an object with a name and a class. This is useful, because R's functions know automatically how to behave differently on different objects depending on their class.

To illustrate the point, let's create two objects, each with a different class and see how the function **summarise** behaves differently, depending on the type. This behaviour is *polymorphism* (Matloff, 2011):

```
# Create a character and a vector object
char_obj <- c("red", "blue", "red", "green")
num_obj <- c(1, 4, 2, 532.1)

# Summary of each object
summary(char_obj)
```

---

[1]We can get the right answer in Python, by typing the following: `import numpy;` `a=numpy.array([1,2,3]); b=numpy.array([9,8,6]); a+b`.

```
##      Length       Class       Mode
##           4   character   character
```

```
summary(num_obj)
```

```
##      Min. 1st Qu.  Median    Mean 3rd Qu.    Max.
##         1       2       3     135     136     532
```

```
# Summary of a factor object
fac_obj <- factor(char_obj)
summary(fac_obj)
```

```
## blue green   red
##    1     1     2
```

In the example above, the output from **summary** for the numeric object **num_obj** was very different from that of the character vector **char_obj**. Note that although the same information was contained in **fac_obj** (a factor), the output from **summary** changes again.

Note that objects can be called almost anything in R with the exceptions of names beginning with a number or containing operator symbols such as -, ^ and brackets. It is good practice to think about what the purpose of an object is before naming it: using clear and concise names can save you a huge amount of time in the long run.

13.3 Subsetting in R

R has powerful, concise and (over time) intuitive methods for taking subsets of data. Using the SimpleWorld example we loaded in *Data preparation*, let's explore the **ind** object in more detail, to see how we can select the parts of an object we are most interested in. As before, we need to load the data:

```
ind <- read.csv("data/SimpleWorld/ind.csv")
```

Now, it is easy from within R to call a single individual (e.g. individual 3) using the square bracket notation:

```
ind[3,]
```

```
##   id age sex
## 3  3  35   m
```

The above example takes a subset of `ind` all elements present on the 3rd row: for a 2 dimensional table, anything to the left of the comma refers to rows and anything to the right refers to columns. Note that `ind[2:3,]` and `ind[c(3,5),]` also take subsets of the `ind` object: the square brackets can take *vector* inputs as well as single numbers.

We can also subset by columns: the second dimension. Confusingly, this can be done in four ways, because `ind` is an R `data.frame`[2] and a data frame can behave simultaneously as a list, a matrix and a data frame (only the results of the first are shown):

```
ind$age # data.frame column name notation I
```

```
## [1] 59 54 35 73 49
```

```
# ind[, 2] # matrix notation
# ind["age"] # column name notation II
# ind[[2]] # list notation
# ind[2] # numeric data frame notation
```

It is also possible to subset cells by both rows and columns simultaneously. Let us select query the gender of the 4th individual, as an example (pay attention to the relative location of the comma inside the square brackets):

```
ind[4, 3] # The attribute of the 4th individual in column 3
```

```
## [1] f
## Levels: f m
```

A commonly used trick in R that helps with the analysis of individual level data is to subset a data frame based on one or more of its variables. Let's subset first all females in our dataset and then all females over 50:

---

[2]This can be ascertained by typing `class(ind)`. It is useful to know the class of different R objects, so make good use of the `class()` function.

```
ind[ind$sex == "f", ]
```

```
##   id age sex
## 4  4  73   f
## 5  5  49   f
```

```
ind[ind$sex == "f" & ind$age > 50, ]
```

```
##   id age sex
## 4  4  73   f
```

In the above code, R uses relational operators of equality (==) and inequality (>) which can be used in combination using the & symbol. This works because, as well as integer numbers, one can also place *boolean* variables into square brackets: `ind$sex == "f"` returns a binary vector consisting solely of TRUE and FALSE values.[3]

13.4 Further R resources

The above tutorial should provide a sufficient grounding in R for beginners to understand the practical examples in the book. However, R is a deep language and there is much else to learn that will be of benefit to your modelling skills. There are many excellent books and tutorials that teach the fundamentals of R for a variety of applications. The following resources, in ascending order of difficulty, are highly recommended:

- *Introduction to visualising spatial data in R* (Lovelace and Cheshire 2014) provides an introductory tutorial on handling spatial data in R, including the administrative zone data which often form the building blocks of spatial microsimulation models in R.
- *Introduction to scientific programming and simulation using R* (Jones et al. 2014) is an accessible and highly practical course that will form a solid foundation for a range of modelling applications, including spatial microsimulation.
- *An Introduction to R* (Venables et al. 2014) is the foundational introductory R manual, written by the software's core developers and is available on-line for free. It is terse and covers some advanced topics, but provides a useful reference on the fundamentals of R as a language.

---

[3]Thus, yet another way to invoke the 2nd column of ind is the following: `ind[c(F, T, F)]`! Here, T and F are shorthand for "TRUE" and "FALSE" respectively.

- *Advanced R* (Wickham 2014) (`http://www.crcpress.com/product/isbn/9781466586963`) delves into the heart of the R language. It contains many advanced topics, but the introductory chapters are straightforward. Browsing some of the pages on Advanced R's website (`http://adv-r.had.co.nz/`) and trying to answer the questions that open each chapter provides a taste of the book and an excellent way of testing and improving one's understanding of the R language.

Glossary

- **Algorithm**: a series of computer commands executed in a specific order for a pre-defined purpose. Algorithms process input data and produce outputs.

- **Constraints** are variables used to estimate the number (or weight) of individuals in each zone. Also referred to by the longer name of **constraint variable**. We tend to use the term **linking variable** in this book because they *link* aggregate and individual level datasets.

- **Combinatorial optimisation** is an approach to spatial microsimulation that generates spatial microdata by randomly selecting individuals from a survey dataset and measuring the fit between the simulated output and the constraint variables. If the fit improves after any particular change, the change is kept. Williamson (2007) provides a practical user manual. Harland (2013) provides a practical demonstration of the method implemented in the Java-based Flexible Modelling Framework (FMF).

- **Data frame**: a type of object (formally referred to as a class) in R, data frames are square tables composed of rows and columns of information. As with many things in R, the best way to understand data frames is to create them and experiment. The following creates a data frame with two variables: name and height:

  Note that each new variable is entered using the command c() this is how R creates objects with the *vector* data class, a one dimensional matrix — and that text data must be entered in quote marks.

- **Deterministic reweighting** is an approach to generating spatial microdata that allocates fractional weights to individuals based on how representative they are of the target area. It differs from combinatorial optimisation approaches in that it requires no random numbers. The most frequently used method of deterministic reweighting is IPF.

- **For loops** are instructions that tell the computer to run a certain set of command repeatedly. for(i in 1:9) print(i), for example will print the value of i 9 times. The best way to further understand for loops is to try them out.

- **Iteration**: one instance of a process that is repeated many times until a predefined end point, often within an *algorithm*.

- **Iterative proportional fitting** (IPF): an iterative process implemented in mathematics and algorithms to find the maximum likelihood of cells that are constrained by multiple sets of marginal totals. To make this abstract definition even more confusing, there are multiple terms which refer to the process, including 'biproportional fitting' and 'matrix raking'. In plain English, IPF in the context of spatial microsimulation can be defined as *a statistical technique for allocating weights to individuals depending on how representative they are of different zones.* IPF is a type of deterministic reweighting, meaning that random numbers are not needed to generate the result and that the output weights are real (not integer) numbers.

- A **linking variable** is a variable that is shared between individual and aggregate level data. Common examples include age and sex (the linking variables used in the SimpleWorld example): questions that are commonly asked in all kinds of survey. Linking variables are also referred to as **constraint variables** because they *constrain* the weights for individuals in each zone.

- **Microdata** is the non-geographical individual level dataset from which synthetic **spatial microdata** are usually derived. This sample of the target population has also been labelled as the 'seed' (e.g. Barthelemy and Toint, 2012) and simply the 'survey data' in the academic literature. The term microdata is used in this book for its brevity and semantic link to spatial microdata.

- The **population base** roughly equivalent to the 'target population', used by statisticians to describe the population about whom they wish to draw conclusions based on a 'sample population'. The sample population, is the group of individuals who we have individual level data for. In aggregate level data, the **population base** is the complete set of individuals represented by the counts. A common example is the variable "Hours worked": only people aged 16 to 74 are generally thought of as working, so, if there is no NA (no answer) category, the population base is not the same as the total population of an area. A common problem faced by people using spatial microsimulation methods is incompatibility between aggregate constraints that use different
population bases.

- **Population synthesis** is the process of converting input data (generally non-geographical **microda** and geographically aggregated **constraint variables**) into **spatial microdata**.

- **Spatial microdata** is the name given to individual level data allocated to mutually exclusive geographical zones (see Figure 5.1 above). Spatial microdata is useful because it provides multi level information, about the relationships between individuals and where they live. However, due to the high costs of large surveys and restrictions on the release of geocoded individual level data, spatial microdata is rarely available to researchers. To overcome this issue, most spatial microsimulation research employs methods of **population synthesis** to generate representative spatial microdata.

- **Spatial microsimulation** is the name given to an approach to modelling that comprises a series of techniques that generate, analyse and model individual level data allocated to small administrative zones. Spatial microsimulation is an approach for understanding processes that operate on individual and geographical levels.

- A **weight matrix** is a 2 dimensional array that links non-spatial *microdata* to geographical zones. Each row in the weight matrix represents an individual and each column represents a zone. Thus, in R notation, the weight matrix `w` has dimensions of `nrow(ind)` rows by `nrow(cons)` where `ind` and `cons` are the microdata and constraints respectively. The value of `w[i,j]` represents the extent to which individual `i` is representative of zone `j`. `sum(w)` is the total population of the study area. The weight matrix is an efficient way of storing spatial microdata because it does not require a new row for every additional individual in the study area. For a weight matrix to be converted into spatial microdata, all the values of the wieghts must be integers. The conversion of an integer weight matrix into an integer weight matrix is known as *integerisation*.

Bibliography

Avram, S., Figari, F., Leventi, C., Levy, H., Navicke, J., Matsaganis, M., Militaru, E., Paulus, A., Rastrigina, O., Sutherland, H., 2012. The distributional effects of fiscal consolidation in 9 EU countries. Social Situation Observatory Research Note 01.

Axhausen, K.K.W., Müller, K., Axhausen, K.K.W., 2011. Population synthesis for microsimulation: State of the art, in: Annual Meeting of the Transportation Research Board, IVT Working Paper. Swiss Federal Institute of Technology Zurich.

Baroni, E., Richiardi, M., 2007. Orcutt's Vision, 50 years on. October.

Barthelemy, J., Toint, P., 2015. A Stochastic and Flexible Activity Based Model for Large Population. Application to Belgium. Journal of Artificial Societies and Social Simulation 18, 15.

Barthelemy, J., Toint, P.L., 2013. Synthetic Population Generation Without a Sample. Transportation Science 47, 266–279. doi:10.1287/trsc.1120.0408 (http://dx.doi.org/10.1287/trsc.1120.0408)

Barthélemy, J., 2014. A parallelized micro-simulation platform for population and mobility behaviour-Application to Belgium (PhD thesis). University de Namur.

Batty, M., 2005. Cities and Complexity, ed.

Castle, C., Crooks, A., 2006. Principles and Concepts of Agent-Based Modelling for Developing Geospatial Simulations 44. doi:ISSN: 1467-1298 (http://dx.doi.org/ISSN:1467-1298)

Chai, T., Draxler, R.R.R., 2014. Root mean square error (RMSE) or mean absolute error (MAE)? – Arguments against avoiding RMSE in the literature. Geoscientific Model Development 7, 1247–1250. doi:10.5194/gmd-7-1247-2014 (http://dx.doi.org/10.5194/gmd-7-1247-2014)

Clarke, M., Holm, E., 1987. Microsimulation methods in spatial analysis and planning. Geografiska Annaler. Series B. Human Geography 69, 145–164.

Deming, W., Stephan, F.F., 1940. On a least squares adjustment of a sampled frequency table when the expected marginal totals are known. The Annals of Mathematical Statistics.

Diez, D.M., Barr, C.D., Cetinkaya-Rundel, M., 2012. OpenIntro statistics, ed. CreateSpace independent publishing platform.

Edwards, K., Clarke, G., 2013. SimObesity: Combinatorial Optimisation (Deterministic) Model. Spatial Microsimulation: A Reference Guide for Users 69–85. doi:10.1007/978-94-007-4623-7 (http://dx.doi.org/10.1007/978-94-007-4623-7)

Edwards, K.L., Clarke, G.P., Thomas, J., Forman, D., 2010. Internal and External Validation of Spatial Microsimulation Models: Small Area Estimates of Adult Obesity. Applied Spatial Analysis and Policy 4, 281–300. doi:10.1007/s12061-010-9056-2 (http://dx.doi.org/10.1007/s12061-010-9056-2)

Gaber, J., 2007. Simulating Planning: SimCity as a Pedagogical Tool. Journal of Planning Education and Research 27, 113–121. doi:10.1177/0739456X07305791 (http://dx.doi.org/10.1177/0739456X07305791)

Grimm, V., Railsback, S.F., 2011. Agent-based and individual-based modeling: a practical introduction, ed. Princeton University Press Princeton, NJ.

Harland, K., 2013. Microsimulation model user guide: flexible modelling framework. National centre for research methods, NCRM working papers. doi:http://eprints.ncrm.ac.uk/3177/2/microsimulation_model.pdf (http://dx.doi.org/http://eprints.ncrm.ac.uk/3177/2/microsimulation/_model.pdf)

Harland, K., Heppenstall, A., Smith, D., Birkin, M., 2012. Creating Realistic Synthetic Populations at Varying Spatial Scales: A Comparative Critique of Population Synthesis Techniques. Journal of Artificial Societies and Social Simulation 15, 1.

Hensher, D.A., 2008. Climate change, enhanced greenhouse gas emissions and passenger transport - What can we do to make a difference? Transportation Research Part D: Transport and Environment 13, 95–111. doi:10.1016/j.trd.2007.12.003 (http://dx.doi.org/10.1016/j.trd.2007.12.003)

Hensher, D.A., 2002. A Systematic Assessment of the Environmental Impacts of Transport Policy. Environmental and Resource Economics 22, 185–217.

Hensher, D.A., Rose, J.M., Greene, W.H., 2015. Applied Choice Analysis, Second Edi. ed. Cambridge University Press, Cambridge.

Hensher, D.A., Ton, T., 2002. TRESIS: A transportation, land use and environmental strategy impact simulator for urban areas. Transportation 29, 439–457.

Hidas, P., 2005. A functional evaluation of the AIMSUN, PARAMICS and VISSIM microsimulation models. Road & Transport Research 14, 45–59.

Hongbin Zhang, G.S., 2002. 35, 701–711.

Hornik, K., 2012. Are There Too Many R Packages? Austrian Journal of Statistics 41, 59–66.

Ince, D.C., Hatton, L., Graham-Cumming, J., 2012. The case for open computer programs. Nature 482, 485–8. doi:10.1038/nature10836 (`http://dx.doi.org/10.1038/nature10836`)

Kabacoff, R., 2011. R in Action, ed. Manning Publications Co.

Knoblauch, K., Maloney, L.T., 2012. Modeling psychophysical data in R, ed. Springer.

Lenormand, M., Deffuant, G., 2012. Generating a synthetic population of individuals in households: Sample-free vs sample-based methods. arXiv preprint arXiv:1208.6403.

Lima, A., Rossi, L., Musolesi, M., 2014. Coding Together at Scale: GitHub as a Collaborative Social Network. arXiv preprint arXiv:1407.2535.

Lovelace, R., Ballas, D., 2013. "Truncate, replicate, sample": A method for creating integer weights for spatial microsimulation. Computers, Environment and Urban Systems 41, 1–11. doi:10.1016/j.compenvurbsys.2013.03.004 (`http://dx.doi.org/10.1016/j.compenvurbsys.2013.03.004`)

Lovelace, R., Ballas, D., Birkin, M.M., Leeuwen, E. van, Ballas, D., Leeuwen, E. van, Birkin, M.M., 2015. Evaluating the performance of Iterative Proportional Fitting for spatial microsimulation: new tests for an established technique. Journal of Artificial Societies and Social Simulation 18, 21.

Lovelace, R., Ballas, D., Watson, M., 2014. A spatial microsimulation approach for the analysis of commuter patterns: from individual to regional levels. Journal of Transport Geography 34, 282–296. doi:http://dx.doi.org/10.1016/j.jtrangeo.2013.07.008 (`http://dx.doi.org/http://dx.doi.org/10.1016/j.jtrangeo.2013.07.008`)

Lovelace, R., Philips, I., 2014. The "oil vulnerability" of commuter patterns: A case study from Yorkshire and the Humber, UK. Geoforum 51, 169–182. doi:http://dx.doi.org/10.1016/j.geoforum.2013.11.005 (`http://dx.doi.org/http://dx.doi.org/10.1016/j.geoforum.2013.11.005`)

Lucas, K., 2012. Transport and social exclusion: Where are we now? Transport Policy 20, 105–113. doi:10.1016/j.tranpol.2012.01.013 (`http://dx.doi.org/10.1016/j.tranpol.2012.01.013`)

M., B., 1991. Evaluation de la demande en trafic : quelques méthodes de distribution. Annales de la Société Scientifique de Bruxelles 105, 17–66.

MacKay, D., 2008. Sustainable energy-without the hot air, ed. UIT Cambridge.

Mannion, O., Lay-Yee, R., Wrapson, W., Davis, P., Pearson, J., 2012. JAMSIM:

A microsimulation modelling policy tool. Journal of Artificial Societies and Social Simulation 15, 8.

Matloff, N., 2011. The Art of R Programming, ed. No Starch Press.

McNutt, M., 2014. Journals unite for reproducibility. Science 346, 679. doi:10.1126/science.aaa1724 (http://dx.doi.org/10.1126/science.aaa1724)

Meindl, B., Templ, M., Alfons, A., Kowarik, A., Mathieu Ribatet, 2015. simPop: Simulation of synthetic populations for survey data considering auxiliary information, ed.

Norman, P., 1999. Putting Iterative Proportional Fitting (IPF) on the Researcher's Desk (No. October). School of Geography, University of Leeds.

Orcutt, G.H.G., 1957. A new type of socio-economic system. The Review of Economics and Statistics 39, 116–123.

Owen, W., 2006. The r guide. Comprehensive R Archive Network.

Peng, R.D., Dominici, F., Zeger, S.L., 2006. Reproducible epidemiologic research. American journal of epidemiology 163, 783–9. doi:10.1093/aje/kwj093 (http://dx.doi.org/10.1093/aje/kwj093)

Popper, K., 1959. The Logic of scientific discovery, ed. Hutchinson.

Pritchard, D.R., Miller, E.J., 2012. Advances in population synthesis: fitting many attributes per agent and fitting to household and person margins simultaneously. Transportation 39, 685–704. doi:10.1007/s11116-011-9367-4 (http://dx.doi.org/10.1007/s11116-011-9367-4)

Rahman, A., 2009. Small Area Estimation Through Spatial Microsimulation Models, in: 2nd International Microsimulation Association Conference. Ottawa; Canada.

Rao, J.N.K., 2003. Small area estimation, ed. John Wiley & Sons.

Rey, S.J., 2014. Open regional science. The Annals of Regional Science 52, 825–837. doi:10.1007/s00168-014-0611-7 (http://dx.doi.org/10.1007/s00168-014-0611-7)

Sutherland, H., Figari, F., 2013. EUROMOD: the European Union tax-benefit microsimulation model. International Journal of Microsimulation 6, 4–26.

Tanton, R., Vidyattama, Y., Nepal, B., McNamara, J., 2011. Small area estimation using a reweighting algorithm. Journal of the Royal Statistical Society. Series A 174, 931–951. doi:10.1111/j.1467-985X.2011.00690.x (http://dx.doi.org/10.1111/j.1467-985X.2011.00690.x)

Thiele, J., 2014. R Marries NetLogo: Introduction to the RNetLogo Package. Journal of Statistical 58, 1–41.

Thiele, J., Kurth, W., Grimm, V., 2014. Facilitating Parameter Estimation and Sensitivity Analysis of Agent-Based Models: A Cookbook Using NetLogo and'R'. Journal of Artificial Societies and ... 17.

Thiele, J.C., Kurth, W., Grimm, V., 2012. Agent-Based Modelling: Tools for Linking NetLogo And R Journal of Artificial Societies and Social Simulation 15, 8.

Tomintz, M.N.M., Clarke, G.P., Rigby, J.E.J., 2008. The geography of smoking in Leeds: estimating individual smoking rates and the implications for the location of stop smoking services. Area 40, 341–353.

Urbanek, S., 2013. rJava: Low-level R to Java interface, ed.

Verzani, J., 2011. Getting started with rStudio, ed. " O'Reilly Media, Inc."

Voas, D., Williamson, P., 2001. Evaluating Goodness-of-Fit Measures for Synthetic Microdata. Geographical and Environmental Modelling 5, 177–200. doi:10.1080/13615930120086078 (http://dx.doi.org/10.1080/13615930120086078)

Wegener, M., 2011. From Macro to Micro—How Much Micro Is Too Much? Transport Reviews 31, 161–177.

Wickham, H., 2014a. Advanced R, ed. CRC Press.

Wickham, H., 2014b. Tidy data. The Journal of Statistical Software 14.

Williamson, P., 2007. CO Instruction Manual: Working Paper 2007/1 (v. 07.06.25), Population (English edition). University of Liverpool.

Index